'In climate discussions, people tend to fall into one of two camps: the people who know a lot and the people who feel a lot. It's rare to find people who both know exactly what they're talking about and are able to eloquently express the sadness, fury and fear that comes with the climate emergency. This book is written by those people. We're lucky to have it.'

ZADIE SMITH, AUTHOR

'There is a habit of people saying, "science should not be political", but science changes our world and our possibilities, so it has no choice but to be political. This collection of essays is a vital read in a world so often obscured by those whose only interest is money.'

ROBIN INCE, COMEDIAN, WRITER AND BROADCASTER

'Eventually, in today's agonizing campaign to address the Climate Emergency, good science will win out over misinformation and denial. In the meantime, more and more scientists have chosen to put themselves on the frontline by taking non-violent direct action. *Scientists on Survival* highlights the eloquent testimonies of some of these inspiring campaigners. We must all continue to trust in science – and scientists like these who so courageously walk the talk.'

JONATHON PORRITT, WRITER AND CAMPAIGNER

'Scientists are usually restrained by academic dogma but here they talk from the heart about what the planetary emergency means to them as individuals. This collection of essays is deeply moving, inspiring and profoundly important.'

MATTHEW TODD, WRITER

"[*The authors of this book*] offer a window into the motivations and coping strategies of a group of scientists willing to forgo "successful" careers, endure media attacks, and face ridicule from colleagues. To me, their reasoned activism represents the relinquished conscience of a scientific community that all too often favours short-term self-preservation over integrity and long-term sustainability. I firmly believe that bad things occur when good people remain silent. I hope this collection of essays will serve as a catalyst for widespread, vocal and informed action from scientists of all disciplines and backgrounds.'

KEVIN ANDERSON, PROFESSOR OF ENERGY AND CLIMATE CHANGE

'At a time when it's often hard to keep hope for a better future alive, these stories from scientists who are choosing to take action are deeply inspiring. They should be a wake-up call for all of us.'

NATASHA WALTER, AUTHOR, JOURNALIST AND CHARITY FOUNDER

'[*Scientists*] permit us to think intelligently about our role as guardians of our fragile planet and how to make informed decisions that minimize harm. *Scientists on Survival* is packed with their wisdom, insights and inspiration.'

DAME VIRGINIA MCKENNA & WILL TRAVERS OBE, BORN FREE FOUNDATION CO-FOUNDERS

SCIENTISTS ON SURVIVAL

PERSONAL STORIES OF CLIMATE ACTION

Michael O'Mara Books Limited

First published in Great Britain in 2025 by
Michael O'Mara Books Limited
9 Lion Yard
Tremadoc Road
London SW4 7NQ

EU representative:
Authorised Rep Compliance Ltd
Ground Floor
71 Baggot Street Lower
Dublin D02 P593
Ireland

A CIP catalogue record for this book is available from the British Library.

This product is made of material from well-managed, FSC®-certified forests and other controlled sources. The manufacturing processes conform to the environmental regulations of the country of origin.

For further information see www.mombooks.com/about/sustainability-climate-focus

Report any safety issues to product.safety@mombooks.com
and see www.mombooks.com/contact/product-safety

ISBN: 978-1-78929-732-4 in paperback print format
ISBN: 978-1-78929-771-3 in ebook format

2 3 4 5 6 7 8 9 10

Cover design by Natasha Le Coultre
Front cover photograph by Crispin Hughes
Illustrations by Lucy Hogarth
Typeset by Barbara Ward

Printed and bound by CPI Group (UK) Ltd, Croydon, CR0 4YY

www.mombooks.com

This book is dedicated to all who act
to protect life on Earth …

especially to people around the world facing
serious consequences for their courage.

CONTENTS

FOREWORD

by Chris Packham,
Wildlife TV presenter and campaigner

For me science is the art of understanding truth and beauty.

And the truth is undeniable: life on Earth is in desperate peril.

Our world is burning, melting, drowning and degrading in ways we can simply no longer ignore. Scientists have dutifully sounded the alarm for decades, yet those warnings have all but fallen on deaf ears. Powerful vested interests — fossil fuels, industrial agriculture and chemical giants — have obscured the facts and silenced the voices of those truth tellers. And despite all the overwhelming evidence, governments continue to stand still, or at least are too slow to act, as our planet continues to sicken.

A few years ago, I met a group of scientists who had decided to resist this dangerous inertia. Like many concerned citizens, they march, they sign petitions, they speak out. But they've also realized that these actions, while important, may just not be enough. In the face of a crisis this critical, we need more powerful methods to drive the changes we so desperately need. So these scientists aren't just raising awareness – they are challenging the status quo. Some have even gone so far as to peacefully break the law, understanding that the stakes are far too high to allow bad business to continue as usual.

In 2023, I stood with one hundred of these scientists at the UK Parliament, urging MPs to recognize the dangers of new oil and gas exploration. Shortly after, following the release of a devastating report on biodiversity decline in the UK, we gathered once more – this time in front of the Department for Environment, Food & Rural Affairs (Defra), alongside forty environmental groups. We blocked a road, shared stories and listened to gut-wrenching yet ultimately hopeful testimonies from researchers, wildlife advocates and activists, all united by a common cause: to defend the fragile ecosystems that sustain us.

More recently, I joined them in a two-day peaceful occupation of a new energy exhibit at the Science Museum in London, sponsored – ironically – by a major coal producer. I have had the privilege of meeting and working with many of the scientists whose voices fill the pages of this book. I know them to be driven not only by their deep love of the natural world but also by their compassion, courage and unwavering dedication.

From the astrophysics student who can no longer bear to gaze at the stars, to the professor who once advised governments, these people have put their heads above the parapet to try something different. Each author, like each chapter, is unique. But what unites them is not just a shared understanding of the science, but a profound personal connection to all living creatures.

This is not a book about the science of climate breakdown or biodiversity loss. It is a collection of human stories told through a fierce love for our natural world. I am deeply moved by their accounts, not only because I share their fears and sense of urgency, but because they also fill me with the hope and the energy to carry on fighting to protect the one thing I love most: life on Earth, our one beautiful and precious home.

INTRODUCTION

by Dr Abi Perrin

Who are Scientists for Extinction Rebellion?

Among us are astrophysicists, biochemists, ecologists, engineers, geologists, immunologists, psychologists, zoologists and many more. All of us have at some point studied a scientific discipline at university and most of us have spent much of our working lives in a scientific profession. Many of us work or have worked as researchers in laboratories or doing practical science 'in the field'. Some of us have or have had roles in science education, in medical settings or in industry. Some of us have spent decades associated with prestigious institutions, some of us are students embarking on a scientific career in a rapidly changing world, some of us have walked away from traditional professions to turn our focus to creating change by other means. As well as being scientists, we might be artists, carers, community organizers, educators, gardeners, grandparents, performers, parents and much more.

What unites us is our fascination with and love for the world around us, and the incredible diversity of people and species which are part of it. We share a desire not only to understand our world but to protect it, too.

We know that the world we love is existentially threatened, and that scientists just 'doing the science' hasn't been enough to change the perilous course that all living beings are on. We agree that we need to do more to stop the harm that is being caused by 'business as usual' amid a climate and ecological crisis. We believe that it is possible to create better systems and a fairer future. We recognize that an emergency situation calls for emergency action.

How did we get here?

If you are reading this book we probably don't need to tell you that life on Earth is in serious danger. We are experiencing the symptoms of unsustainable, extractive and exploitative practices: human-driven planetary heating, pollution and ecocide are causing and accelerating the extinction of countless species, the erasure of irreplaceable ecosystems, and the suffering, displacement and deaths of millions of people. There is a glaring disparity and inequality between the people and places that have profited most from destructive industries and those who are currently facing the devastating daily realities of climate and ecological breakdown. We know that the combined impacts of the climate and biodiversity crises will continue to escalate ... but collectively we have a choice about how bad, how rapid and how unjust we allow those impacts to be.

There are many books and reports already out there that lay this situation bare. A selection of those we recommend are listed in the Suggested Further Reading section of this book.

Our collective theory of change

Growing up, we were each taught that science generates evidence, which in turn informs policy and creates positive societal and global change. But now we recognize that this social contract has been broken – or perhaps it never really existed. Scientists like us have been sounding the alarm for decades, but those warnings have been ignored, dismissed or suppressed. Too often have the vested interests of some of the world's wealthiest and most powerful industries (e.g. fossil fuel and agrochemical) and individuals (certain lobbyists and politicians) muddied the waters and delayed action to address global heating, biodiversity destruction and the associated harms to people. It can feel overwhelming, frightening and distressing to grapple with such existential threats, so as a result there are many psychological incentives to look away and to act as if we aren't confronted with an immediate danger. But at this critical point, we cannot afford to look away.

What does it mean for us as scientists, if our warnings haven't protected our climate, ecosystems and communities? We believe it means that scientists must change tactics. We cannot continue to rely solely on doing the research and publishing reports. We need to go beyond the 'traditional', comfortable methods of writing letters and signing petitions to communicate vital messages and advocate for emergency action. We acknowledge that scientists occupy a privileged and trusted position in our society, and we feel a responsibility to use that to press for urgent change.

We can apply that pressure in so many ways – through teaching, communication, campaigning and more – but we know from history and social science that collective nonviolent civil disobedience is a hugely powerful tool capable of accelerating societal change. Based on that knowledge, each of us has decided to support or join peaceful actions in protest against the continued assaults on living beings (human and otherwise) and in demand of equitable, regenerative, democratic and inclusive alternatives.

A call to action

What would it look like if we expanded our conception of what a scientist needs to be in the twenty-first century? What if every scientist were to stand up for science, joining thousands of their colleagues to demand the systemic changes we desperately need? What if scientists learned from, immersed themselves in and became an integral part of social movements fighting for environmental justice? With the stakes so high, the window of opportunity to change course closing so rapidly, and the opportunity – and responsibility – for scientists to make a difference, what's stopping us?

A guide to this book

This is not another book about the science of the climate, ecological and inequality crises. It is something a little different. We are attempting to illustrate the scale of the emergency, convey our motivations and explain our actions by telling our stories. We have learned from many years of trying to communicate through graphs and data that it's not a lack of available scientific knowledge that holds back action, rather a problem of emotional connection with the enormity of the issues, uncertainty about where to direct our focus, and fear of what we might experience if we embark on certain courses of action. So this book is not an encyclopaedia of hard-hitting stats, but instead a collection of short personal stories from scientists. We have tried to break the mould of traditional science communication and instead show the ideas that motivate us through accounts of our perspectives and experiences as both scientists and activists.

The chapters that follow can be read in any order. We hope that you find stories that resonate with you, challenge you, and inspire you.

Note: All of the authors are writing from their individual perspective of being a scientist–activist as part of the environmental movement in the UK. We acknowledge that the diversity of our group, and consequently the authors, is not representative of the demographics of the UK. Specifically Asian, Black and minority ethnic groups, as well as working class and disabled people are under-represented. This book also does not include the perspectives of scientists from the Global South, where the worst of the current impacts of the climate, biodiversity and inequality crises are being felt, and where activists and Land Defenders face significantly greater oppression and violence.

PART 1

SCIENTISTS' PERSPECTIVES

'WALKING IN WATER'

HUMANS: THE CAUSE OF AND SOLUTION TO ALL THE PLANET'S PROBLEMS

Stuart Capstick, PhD

Stuart is an Honorary Senior Research Fellow at Cardiff University, former lead for the Tyndall Centre for Climate Change Research at Cardiff University, and former Deputy Director of the Centre for Climate Change and Social Transformations, which he helped establish in 2019. He has jointly led work for the United Nations Environment Programme on lifestyle change to address the climate crisis, written a chapter for Greta Thunberg's The Climate Book *on personal action for social change, and been an author on more than fifty peer-reviewed journal articles. In late 2023 he moved from academic research to take up a post with a conservation charity.*

I've often wondered what it is that moves someone from being vaguely concerned about the environment, to that visceral grasp of the true scale, horror and injustice of the emergency heading our way. The feeling deep in your heart and stomach that this really is it. That the indescribable beauty and vitality of the natural world is collapsing around us and is about to take human civilization with it.

I'd like to be able to tell you that the start of my understanding came from an encounter with a passionate activist deep in some protest crowd, or when I discovered a new species of frog halfway up a mountain. But the faintly embarrassing reality is that it was Al Gore's slide show presentation in the film *An Inconvenient Truth* that changed things for me. Having slunk off to watch the film alone in a cinema one weekday morning when I should have been at work, I walked home with facts, statistics and warnings rattling around my head and knew that something was different now. Soon after, I started a PhD in climate change psychology and for the next fifteen years spent my days researching, publishing, teaching, speaking and collaborating with a wide range of climate change academics and practitioners. I've learned a lot and have had the chance to work with some excellent people. I'm better equipped than ever with knowledge accumulated from the vast and growing literature. I've tried hard to look at the issues from different angles and to remain open-minded, to always question while remaining faithful to the science. And I feel more vulnerable, confused, heartbroken and afraid than ever. Why do we know but not act? Why do we participate so willingly in our own downfall? Is it the sad truth that we'll only know what we've got once it's gone?[1]

1 I've used the first-person plural 'we' and 'us' throughout, as a shorthand to talk about people in general or the activist community. From the perspective of climate justice, there are very large disparities between who is causing emissions and who is vulnerable to impacts, who has power and who does not, who is 'we' and who is 'they'. These words should be read with this in mind.

Scientific work on the environment by necessity spans a wide range of disciplines – from atmospheric physics to economics and ecology. But I went into my field of study because of the belief that how we think and behave is at the root of the climate crisis and how it should be addressed – and my experiences and research over the years have only heightened that conviction.

This doesn't mean that everything comes down to individual consumer choice – a narrow idea that has taken a battering in recent years from environmental campaigners as well as academic researchers. But it does mean that the essence of what it is to be human permeates all aspects of the enormous mess we're in. Our blinkered outlook can be seen every day in millions of streets and towns around the world, as people sit in their parked cars, engines idling while they tap at their phones, mindlessly pumping CO_2 and dirt into the air. Consumer society and its over-abundance of 'stuff' relies upon someone to buy that SUV for the status it conveys, or to click on a cheap flight in search of the dream of travel. The inequalities and power imbalances that have been a constant struggle throughout history are amplified through a crisis in which the haves blithely inflict harm upon the have-nots and the not-yet-born. The ultimate bogeyman of our time – the capitalist system itself and the politics that accompany it – is in the end entirely our own invention. It grows and mutates with a life of its own, but each of its dials and levers are held by human hands and based on the insatiable promise of more.

Perhaps it's because we are so much a part of the problem that we find it so hard to see a way out. A fish, so they say, can't understand the concept of water, so completely is it immersed in this substance that is all it's ever known. Yes, if pressed, most of us can agree that things don't look good and something has to give. Plenty of research studies, including ones I've worked on, show that people recognize the reality and importance of environmental problems. But on a day-to-day basis, how many of us really carry this around with us, allow the vastness and the urgency of it to influence how we see the world and the way we live our lives? How many of us would *want* to? Instead we get up, eat breakfast, go to work, watch TV, get drunk, laugh and cry, go to sleep and start all over again, most of us barely acknowledging the reality of the climate and nature emergency unless something makes it impossible to avoid.

There are reasons to be found both in our brains and our societies for this. Some have argued that we are simply wired in the wrong way, that we are just odd-looking apes who hurried down from the trees still in possession of our ancient ways of thinking and without knowing what we were getting ourselves into. Psychologists point to various biases and quirks in how we interpret the world that make dealing with an issue like climate change especially hard. Many of these involve a reliance on emotion over logical reasoning – evolved no doubt to keep us out of trouble when predators were hiding in nearby caves, but of less use when we're dealing with the complex and drawn-out connections between our societies and the warming of the planet. If it doesn't *feel* like things are all that serious – and let's be honest, if you look out the window right now, you can't *see* the natural world unravelling before your eyes – then maybe all those unnerving computer models have got things wrong.

There's something inherently social to the way we are responding to all this too – the ability to conspire together in turning a blind eye – something that the scholar Kari Marie Norgaard has termed socially-organized denial. This is the way in which everything from how we talk about the weather to the constant distractions and amusements of modern life keep things seeming normal and unquestioned, despite all the evidence that our resource-hungry societies are built on a house of cards. It's in this context that the climate activist Matthew Todd has argued that even as we receive ever-starker warnings from scientists and the communities most affected by the climate emergency, they have become now a kind of white noise, hissing dimly in the background. We are used to it all now; we know but, well, that's just how it is. Shrug emoji. Move on.

At the other end of the spectrum, for those of us who have obsessed over these matters, the situation can seem so terrifying that it's hard to believe we are going to take action in time to avert outright catastrophe. The siren song of doom-ism is never too far away, promising a grim relief through finally giving in to despair and resignation at the enormity of it all. Personally, I'm one of life's Eeyores and find it hard to accept that any good news – whether that's the rapid growth in renewable energy, sincere efforts by some parts of industry to do better, or hopeful and creative ideas for living differently – is really doing much to halt the downward slide.

But if the human dimension to the climate crisis is what has led us to this point, here also are the foundations for a change of direction. The climate crisis feeds on our being selfish, short-sighted, thoughtless and dishonest, but we also have the capacity to defy that bleak picture. Although some of the most powerful people in the world apparently are uninterested in the wider interests of humanity, the vast majority of us are not implacable devils: in fact, we have a miraculous ability to feel concern for people we'll never meet or who don't even exist yet, but who still matter. Most of us want to be part of making the world a better place and will readily help out a friend or neighbour in need. Research in social psychology is clear that deep in our psyche we hold a set of ethics that emphasizes care and responsibility for others, a desire for peace and social justice, and compassion and empathy towards the natural world. In the literature these are termed 'self-transcendent values' – stressing the very human ability to step outside our own narrow concerns and act in service of some higher cause. A long-standing body of research shows these values to be woven through all human cultures. Studies are also clear that people who prioritize them tend to be happier as well as kinder.

It seems to me that the new wave of climate activism that arose in the late 2010s embodied that defiance and that kindness, encapsulated by the maxim often used in email sign-offs or social media posts: 'Love and Rage'. Suddenly here was a demand not just to 'tell the truth' but also to act as though the truth is real. It's something that shouldn't sound the least bit controversial, but was nevertheless a radical and rebellious desire to break the ongoing state of stupefaction.

During the first of Extinction Rebellion's demonstrations, the police and mainstream politicians didn't know what to do with this unruly energy, nor had there yet been that series of spiteful Home Secretaries determined to crush the life out of all protest and dissent. For once, it really seemed like there was a critical mass of people able to push important issues onto the agenda. Through being on the streets and refusing to be ignored, we were trying in our own clumsy and faltering way – though make no mistake, mistakes were made – to act as though the truth is real. And amazingly, over the months that followed, it worked: the protests by Extinction Rebellion alongside the Greta Thunberg-inspired school strikes measurably changed public discourse and attention in a way that the science on its own – so important, but so dry, so impenetrable, so full of caveats – never could. The first time I was arrested and placed in a cell for joining one of those brief disruptions of the status quo, I was surprised in the middle of the night to feel a strange and gentle calm come over me; I still can't really put my finger on it, but I think it was because some sense of self-respect had bubbled up through the usual teeth-grinding anxiety and bewilderment about the climate crisis. I'm not saying it was big and clever to end up in that situation (in fact it hadn't really been my intention that day) but I know I was trying to do what felt right at the time.

In the UK at least, much of the earlier momentum unleashed by Extinction Rebellion has been stomped out by good old-fashioned state repression – you can now go to prison for peaceful and only mildly disobedient protest – or has morphed into other things. There are new campaigns and protest movements pushing back against fossil fuel expansion and the tainting of culture by greenwashing, as well as using legal routes to challenge high-carbon developments. New communities have emerged which are aligned with climate activism, including among health professionals, lawyers, businesspeople and scientists. Valuable lessons have been learned: many of us are less naïve these days about power and the way it's exercised; we want to build solidarity with people around the globe, many of whom face far more difficult situations than we're accustomed to in the rich world; and it has become clear that ambitions such as averting a global 1.5°C temperature rise are in effect now consigned to the realms of fantasy.

Nobody knows what happens next. The climate and ecological emergency is guaranteed to get worse because we have yet to see the full consequences of what we have already set in motion and because global emissions remain stubbornly colossal ... but it's up to us how much further down this road we go. We need as many people as possible willing to speak up and take action, to do more than just quietly publish journal articles or do a bit of recycling. You don't have to throw soup on anything. You don't have to glue yourself to things. But if you really do care about all this, please find your own way to act like the truth is real.

WILDLIFE AT THE HEART

Nikki Tagg, PhD

Nikki's route into wildlife conservation was partly academic, and she spent the best part of fifteen years doing research on great ape ecology and conservation in Central Africa. Frustrated at reporting declines in animal populations but not feeling able to halt those declines, she moved to where she is now: heading up the conservation department of a wildlife charity, where she can have a bigger impact as well as a platform to raise awareness and make change. But with such huge pressures working against conservation, Nikki was very receptive to the uprising of the climate movement. Nikki is now attempting to carve out an impactful niche in the climate and biodiversity fight by volunteering for her local Wildlife Trust and Green Party, advocating at home and on social media for all things green, supporting her local Extinction Rebellion group, and being an active member of Scientists for Extinction Rebellion, and by standing up for nature on the streets.

I duck down the footpath and cross the small bridge into the field. It's a cool but clear day, and the budding colours of spring are emerging as if full of questions. My muddied walking boots squelch as I take the well-worn path round the field to the stream, where I'll pause for a bit, as I often do, absorbing the fragile sights and sounds. This is where I'm at my most creative, full of hope, where I can breathe and watch and think, where jumbled thoughts align.

In my childhood, my passion was recognized by people around me as a 'love of animals'. My chosen career in wildlife conservation has over the years been considered admirable and 'the right thing to do', but more commonly as misguided or a distraction. What I dedicate myself to has been deemed by society to be not as important as so many other day-to-day concerns and needs, topics of conversation, or charitable causes; my chosen career path not as valuable or necessary. I have wasted time believing this, being unsure, but now it's all becoming clear.

Recognizing our true values

People in the Global North live in bubbles. We surround ourselves with stuff, immerse ourselves in a human-made world, live without connection to nature, and deny any need for it. For many people, nature can seem peripheral, optional or irrelevant. The pressures of society because of the circumstances we are born into – the lack of choices, poverty, a biased media and so on – often mean that many people's lives are dominated by fashion, money, entertainment, celebrities, gossip, gadgets, haves and have nots. If we could strive to see through these 'junk values' and hone in on what lies at their foundation, we might be surprised at what our true values are – friendship, health, happiness, security. Perhaps nature.

I have done a fair amount of such reflecting, often as I ramble among the wildflowers or through the woodland patch along the edge of a steep gully, where a burst of rustling up ahead alerts me to a deer darting into the undergrowth. I have thought hard, while indulging in the tranquil shapes and hues that are so at risk, and I know now just how strongly I value nature. I value nature for what it is – for the variety, brilliance and the uniqueness of it. For its omniscience and sheer presence. For its magnificence, ingenuity, hilarity. For its pervasiveness and persistence, its right to persist. And for selfish reasons, too. For everything it gives us. Life. Joy. Nature gives me genuine, heartfelt, aching joy. And witnessing the loss of nature at the scale that is underway today is unbearable.

My breath catches in my throat with a wave of panic when I imagine the scale of the loss already suffered. If my grandparents were to walk my well-trodden path, how surprised might they be at the bare and silent natural world of today compared to just a few generations ago? In this developed yet shamefully nature-depleted country, how did it get to this, and how much more will we lose?

As I push on up the hill, I realize that my other values are at risk, too. Unless we act today, our children may not have a future in which they can live free of fear. Honesty and fairness are compromised: millions of people are suffering globally, sacrificed by a wealthy minority which has the power and money to buy the silence of the world's press and win favours from the world's leaders. We are being stripped of our rights to a healthy environment and to the joy and life-giving properties that nature brings.

With this incessant destruction of nature – habitats and the wildlife that live within them – we are compromising our chances of a stable climate and liveable planet. If the first essential step for any chance of a secure future is to stop burning fossil fuels (turning off that proverbial tap), then the second is to protect and restore natural ecosystems in order to soak up what has already overflowed. We cannot undo the damage we have wrought without the vast carbon sequestration properties of the oceans, forests, peatlands, grasslands and so on. This is now widely recognized by nations, governments and society, and numerous pledges have been made, including to restore millions of hectares of forest and protect a third of the world's oceans in the next few years.

Wildlife as our warriors

But there has been a tendency to overlook the ecosystem 'engineers' that precipitate those planet-preserving properties of ecosystems: wildlife.

For many years, the world's decision-makers have told me that my interests are no more than admirable, and that development and economic growth at the cost of the existence of numerous amazing species across the world – from wolves and turtle doves to salamanders and the rare, crocodilian gharials – is a given. I have been told it is unavoidable, even essential. Although it may be a shame, it is nonetheless just the way things must be. Losing wildlife is an inevitable part of the human story.

I know now that it is in fact the very opposite. The chaotic stampede of humanity down this clear-charted road of development and economic growth at the expense of nature is a suicide mission and we urgently need to turn back. As I gaze up into a huge old oak tree where a gentle sweep of wind is brushing its upper branches, and sparrows quarrel in the bushes behind me, I finally feel the strength of my conviction.

Wildlife conservation is quite possibly the most critical endeavour of our time.

For wild animals ensure the balance and function of their habitat homes. They maintain, regenerate and restore ecosystems. When going about their daily lives, animals are hard at work for the planet.

As forest elephants roam in Africa, they trample on smaller plants and seedlings, giving the larger, carbon-storing trees the space to grow. Whales feed at depth and release nutrients as they breathe and rest at the surface, stimulating the production of microscopic marine organisms called phytoplankton. As well as forming the base of all marine food chains, phytoplankton are fantastic at photosynthesis – harnessing the power of sunlight to capture atmospheric carbon dioxide, storing the carbon, and releasing life-giving oxygen. Pangolins dig – to eat ants and termites and to excavate dens to sleep in – and, in so doing, circulate essential nutrients. Healthy soil ecosystems have an unparalleled carbon sequestration role.

Thanks to growing scientific evidence, the list swells: tigers, bison, wildebeest, sea otters; grazing, foraging, trampling, nest-building. Fertilizing the soil, supplying nutrients, enhancing plant dispersal, reducing the frequency of wildfires. Buffering the bad and boosting the good.

Wildlife is our natural ally, unknowing but all-powerful. It is critical to humanity's future. To enable us to divert from the path to annihilation that humanity is set on, we must protect, restore and rewild on a massive global scale.

I watch the minnows' rippling bodies pushing against the current of the stream. A kestrel sweeps overhead, blurring to a speck as it pulls up over the hill, and my heart stills for the enormity of what has already been lost and the vulnerability of this nature on a knife-edge – desperate remnants among the rich and developed, the failing and the blind.

Duty of care

As a wildlife conservationist, I work for the protection of species like tigers, elephants and gorillas. The objective is often to promote harmonious coexistence in landscapes where people and wildlife both strive to eke out an existence. In such situations, wildlife invariably loses. The crux of our conservation work is asking rural-living people in the Global South – people whose livelihoods and lives are often at risk because of the wildlife they share their homelands with – to accept, adapt and care.

We ask a livestock farmer to accept some losses from their herd of cattle to a hungry lion and to not retaliate against the predator. We ask a family to adapt their farming methods and try growing crops that are unpalatable to the elephants that burst out of the nearby reserve during the dry months when nutritious food is scarce. We show groups of schoolchildren awe-inspiring videos of peaceful gorillas resting in a family group in a sun-dappled clearing, hoping to stir empathy through footage of infants playing, softly panting, their wiry hair bristling as they roll and tickle.

But do we do as we ask of others? Those of us who do not live side by side with wildlife nevertheless rely on nature just as much – for the water we drink and air we breathe – but do we make the necessary sacrifices?

Can we accept that our towns and cities cannot sprawl forever, that we need to stop pouring concrete over our green spaces and decimating the habitats of bats, dormice or great crested newts? Is it time to adapt our shocking, age-old habit of using poisons in our agricultural systems and apply alternative methods, to prevent the rapid and deadly decline of the pollinating insects that secure our food supplies? Could we open our minds and hearts and junk-filled bubbles to every one of the wondrous wildlife species that still cling to life on the British Isles and that we live alongside, from badgers to the Adonis Blue butterfly, and begin to *care*?

I soak up a last glance of the gently pulsing stream then turn to stride back across the muddy field, trusting in the nature I leave behind me. I am buoyed by the experiences and sensations all around, terrified by the threats and possible future that looms ahead of us, but impassioned by the fact that the realization that we must care is rising up around the world, as it is in me.

This realization is a massive awakening for me as a wildlife conservationist and imparts an incredible feeling of responsibility. As the default of human society is to undervalue wildlife, I have a duty to my passion, my values, and my understanding of wildlife conservation to help bring nature from the periphery to the forefront of the human story.

We know now, and there is no time to lose.

WALKING IN WATER

Charlie Gardner, PhD

After working with some of the world's rarest birds in Mauritius and spending a decade helping establish new protected areas in Madagascar, Charlie stepped away from the front lines of biodiversity conservation to try to bring about the transformative change to our societies that's needed if any wildlife is to survive climate change. He is an Associate Senior Lecturer at the Durrell Institute for Conservation and Ecology (University of Kent), a spokesperson for Extinction Rebellion and Scientists for Extinction Rebellion, and a leading advocate of activism by scientists. His book What One Person Can Do in a Planetary Emergency *will be published in 2026.*

Sometimes I would feel it at the coast, as I walked through historic villages or nature reserves teeming with migratory birds. Often I'd feel it on the train as it sped me homewards across the flat expanses of the Cambridgeshire Fens. And I would feel it most strongly as I stared into the maps published by Climate Central,[1] that highlight in startling red the great swathes of eastern England that are projected to be under the annual flood level by 2050. But what was 'it'? That was never really clear. It was such a jumble of emotions that I couldn't always tease them apart. Sadness was a big part, a growing grief for the places I loved which would no longer be. Fear was always present, too, and increasingly anger. But mostly what I felt was simple incredulity: how could it be that these places, so solid under my feet, will be gone?

And how on earth could it possibly be that everyone in these places – the farmers who till them, the tourists who love them, and the residents who have made their lives here – knows they will soon be gone, yet carries on with their life as if nothing will ever change?

I love the Norfolk coast and had long harboured a secret dream to walk its length, but looking at those Climate Central maps one day I realized I could do more than just that: I could walk from the centre of historic Cambridge to the mouth of the Wash, round the coast and into the centre of Norwich – more than 180 miles in total – entirely on land that may soon no longer be there. The very thought boggled my mind, and I felt an urge to get to know these lands, underfoot, before it's too late. And having hatched the thought, I realized that by connecting these places on foot and turning a science story – with the aid of a gimmick – into a human one, I could perhaps spark some conversations and break through the deafening silence. If no one was going to talk about the terrifying future that the east of England faces, perhaps they might talk about the bloke in the mask and snorkel.

I plotted a route, picked a date, and fired off some messages

1 Climate Central is an independent organization of scientists, which reports on climate impacts and produces resources such as the sea level rise visualization tool at coastal.climatecentral.org

inviting friends and fellow activists to join me for a stretch of my stroll. But the wonderful thing about being involved in communities of activists is that you know lots of creative, energetic and thoroughly brilliant people willing to work towards anything that inspires them as a good idea. Within days, my plans for a fairly low-key walk with a bit of local media attention had blossomed into a fully-formed project. Chris built a website and set up our social media. Carol called every journalist on my route and got us more local radio and newspaper coverage than I could possibly have dreamed of. Hannah not only helped craft the message, but also designed a beautiful T-shirt, banners and bookmarks to deliver it. Others joined Zoom meetings and shared their ideas and expertise. And dozens would join me en route when, convinced I was never going to make it after not enough training, I finally set off in September 2023.

From Cambridge I walked north for four days along the banks of the Cam and Great Ouse rivers, through England's agricultural heartlands, until I reached my local town of King's Lynn. Four days through a Fenland landscape that had once been a dynamic delta of coastal wetlands until centuries of drainage claimed it for agriculture. Four days across lands that, following that drainage and the resulting oxidation of their rich peat soils (which has caused them to shrink), are now so low-lying that they're already below the level of the rivers flowing through them.

From there it was another four days east along the coast of North Norfolk, through wetlands, farmlands and villages mere inches above the highest tides, before I hit the cliffs of the Cromer Ridge and we temporarily left the lowlands behind. Deposited by a glacier as it retreated at the end of the last ice age, this ridge is made of sands, gravels and clays so young they haven't had a chance to consolidate into stone, and it crumbles at the faintest hint of a winter storm. This is the fastest disappearing coastline in Europe, and as I walked along the beaches in front of villages like Happisburgh and Hemsby I saw they were littered with bits of pipes, cables and other debris from houses that had fallen off the cliff and been swept into the sea.

Finally, after a week of threatened coastlines, I headed inland at Great Yarmouth and was again confronted with the bizarre, scary sight of a vast landscape spreading well below me as I strolled the banks of the river running through it. This was the Norfolk Broads, and it would take another four days to cross them and reach Norwich. I had walked for fifteen days, well over 180 miles, and none of that land I crossed is safe.

We often talk about climate change as a future threat, and to this day you'll hear people saying we must do something 'for our grandchildren', but for the people of Happisburgh and Hemsby, it's nothing quite so abstract. Nor was it for the residents (and car drivers) disrupted by flooding in Wells-next-the-Sea, Blakeney and Brancaster as I walked through them, caused by a small storm surge driven by Storm Agnes. And nor is it for residents of Heacham in West Norfolk, which I passed through on Day Five, where the scale of the threat was brutally apparent.

In Heacham there are holiday homes and caravan parks, as well as parts of the Wild Ken Hill nature reserve, which are protected from the highest winter tides and storms by a shingle bank. Every year the waves lap at the bank and draw the shingle down the beach, so every year the borough council and the government's Environment Agency send out the bulldozers to bank it back up, but the costs of doing so have ballooned recently. The council and Environment Agency have long maintained a policy of 'hold the front line', but as I write that's under review. By the time you read this, the bulldozers may have been stopped, the bank may have been breached, and the irreversible process of relinquishing this land to the sea will have begun.

Elsewhere along my route the future is harder to foretell. Seeing a map of places that will be under the annual flood level by 2050 implies some certainty, the date provides specificity, but for a lot of places marked in red on those maps the reality is a lot more complex. Many of them, in fact, including most of the Fens and the Broads that I strolled through along the elevated banks of the rivers, are under the annual flood level already. They stay dry because of centuries of investment in defences and drainage, but as sea levels rise and storms become ever more ferocious, they'll become harder and harder to defend, until the day we simply can't defend them anymore. And we have no way of knowing when that may be.

Ultimately, we don't know how fast or how high the oceans will rise, because we don't know how bad planetary heating is going to get. Stabilizing the climate requires slashing greenhouse gas emissions by ending the fossil fuel age and reversing the destruction of nature, which in turn requires a complete transformation in the way our economies and societies operate. Since we cannot say when (or even, incredibly, if) this will happen, the Intergovernmental Panel on Climate Change can't say whether we'll get 50 centimetres of sea level rise by the end of the century or closer to 1 metre. Worse, since there are also uncertainties in the way that ice sheets in Greenland and Antarctica are responding to the warming atmosphere, they cannot rule out 'low probability, high impact events' that could add another metre on top of that. How much sea level rise will we see within the anticipated lifetimes of my young nieces and nephews? We simply don't know.

For low-lying communities along my route there is even less certainty, for of course it's not changes to global averages that do the immediate damage, but extreme weather events. On the North Sea coast of Britain that means storm surges, which happen when a big winter storm coincides with a high tide, and the wind and low pressure blow a bulge of water toward the land. Such storm surges have always happened – one on 31 January 1953 killed more than 300 people across eastern Britain, and thousands in the Netherlands – but they've been getting more frequent, and are expected to become more extreme as the climate worsens. When will the next big storm surge hit? We don't know.

My walk opened my eyes to the realities of the climate impacts already happening near my home, at just 1.2°C of heating,[2] and helped me understand the looming threats to places I love. But it also gave me some insights that have helped me rethink how to respond. As a scientist and activist, I thought my role was to ring the alarm about our planetary emergency and warn everyone about the urgent need to do something. That was what my walk was supposed to be about – raising awareness – but almost everyone I spoke to along the way already knew about the threat. Many, perhaps, didn't know quite how dangerously urgent the situation is, but almost everyone accepted the realities of climate change, was worried about what it would mean for them and their homes, and wanted to do something about it.

But while many were proud of the efforts they were making to reduce their energy use at home, and seemed to want to do more, I also felt a great sense of fatalism, a defeatism stemming from a sense of powerlessness in the face of rapacious fossil fuel companies, a government who had just given a green light to new oil fields in the North Sea, and the sheer scale of the task ahead. Everyone agreed that something had to be done, but nobody seemed to feel that they were the ones to do it. They didn't know what they could do or, if they did, they didn't feel it would make any difference, so instead they just silently got on with their lives. As an activist, I realized, I must do more than just ring the fire alarm: I must also detail the way out of

2 In 2021 the IPCC reported that average global temperatures had increased by 1.1°C since the start of the Industrial Revolution. By 2023 this had increased to 1.2°C.

the building, and show people how to find it. Since then my focus has shifted from simply encouraging people to act, to also showing them how.

The other thing that struck me was how willing most people were to hear what I had to say. I've often felt that nobody wants to talk about this planetary emergency, that raising the subject would be intrusive and unwelcome, and so I've held back and usually kept quiet. But my time on Norfolk's beaches wasn't like that at all. Once I approached people and explained why I looked so silly in my mask and snorkel, many seemed happy to finally be having this conversation. Relieved, even, to have the opportunity to share the worries they'd been keeping bottled up because they too thought nobody wanted to talk about them.

It might seem like nobody's listening when we talk of greenhouse gasses, 2050 and 1.5°C, but maybe that's not how best to break the silence. With a relevant story, a human angle and a little creativity, I think we all can find ways to spark those crucial conversations.

WHAT'S STRESSING THIS IMMUNOLOGIST? POLITICAL INDIFFERENCE TO NATURE, DISEASE AND THE CLIMATE CRISIS

Brian Jones, PhD

Brian, a Londoner, studied microbiology at Cardiff University and went on to gain a PhD in tumour immunology from the Welsh National School of Medicine. After post-doctoral training in transplant immunology, he migrated to Hong Kong to set up and run a Clinical Immunology Laboratory at the Queen Mary University Hospital. After many years of teaching, research and clinical work, he and his wife retired to rural Devon. They have been volunteering with various charities, notably the Royal Society for the Prevention of Cruelty to Animals (RSPCA) and the Royal Society for the Protection of Birds (RSPB), and are committed to enjoying and protecting nature. They made many lifestyle changes on returning to the UK, including becoming vegan and getting deeply involved in campaigning and protest.

I have had a good life in science. But now, well into retirement, I wonder how much I have actually achieved in terms of tangible benefit to humanity amidst political failure to address the twin existential crises of biodiversity loss and climate change. I would like to draw on two very different times in my life to illustrate that difficult problems could be resolved if institutions and politicians took notice of robust scientific evidence. My first example involves the arrival of a novel – and deadly – epidemic infectious disease. My second example is the unnecessary, cruel and expensive decimation of a much-loved native British mammal species.

***

In March 2003, the previously unknown infectious disease Severe Acute Respiratory Syndrome, SARS, crossed the border from Guangdong province in China to Hong Kong, where I was working as an immunologist at the Queen Mary University Hospital. When front-line hospital staff – nurses, doctors and orderlies – became seriously ill and many died, my laboratory staff and I realized that we were all likely to be exposed to a new and mysterious infectious agent. We were afraid, but we knew what we had to do – protect ourselves, our patients and the public from a deadly new disease. In spite of our fears – or maybe because of them – there was a strong sense of camaraderie throughout the hospital; all departments came together in a concerted effort to identify the responsible organism, find out where it came from, work out how it caused its deadly symptoms and find ways to stop it from spreading.

Within months of the first cases, we identified the SARS-CoV-1 coronavirus as the cause of this devastating disease. We developed an effective diagnostic test to distinguish SARS from other diseases with similar symptoms, and we introduced treatments to at least partially control the disease.[1] Although we were working under intense pressure, it felt incredibly satisfying to be gaining insight into an unprecedented medical emergency. We were also heartened that our published work would be available to guide the management of future coronavirus epidemics and would hopefully save lives and distress.

Sadly, the spirit of cooperation within and between university and hospital departments experienced over the course of the epidemic dissipated when, rather unexpectedly, SARS did not recur in the winter of 2004. The collaborative approach we had needed to adopt soon returned to a less productive, more competitive baseline. Perhaps the short-lived nature of the SARS epidemic raised false hopes that we'd quickly be able to get a handle on future potential pandemics ... seventeen years later the complacency with which most politicians dealt with SARS-CoV-2 (COVID-19) would suggest as much.

***

My wife and I retired in 2007, and returned to Britain to live in rural Devon. Aware of the threats posed by global heating and the destruction of nature, we occupied ourselves with making our home as sustainable as we could. We fully insulated the

1 We also observed that some patients suffered long-term incapacities which needed continued treatment and care, just as is the case with 'long COVID' seen with the COVID-19 epidemic. We realized that effective vaccines against SARS-CoV-1 should be developed, but the pharmaceutical industry did not take this up. Thankfully they did for SARS-CoV-2 (COVID-19) in 2020.

house, installed solar panels and a storage battery, and changed from liquid petroleum gas to electric heating. We dug a pond and rewilded an area around it. Conservation volunteering with RSPCA and the RSPB kept us occupied for a few years, but in 2013 I found myself once again immersed in science. The reason? Badgers!

Throughout my career as an immunologist, I'd observed and studied many human diseases. I'd seen first-hand how pandemics unfold and what effective (and ineffective) control measures look like. In particular I had learned that stress could affect the immune system very badly; I had studied how that worked and what could be done about it. So when I saw the approach the Department for Environment, Food and Rural Affairs (Defra) was taking to manage bovine tuberculosis (bTB), I was seriously alarmed.

BTB is a devastating disease. Infected cattle suffer shortness of breath, coughs and weakness, and even when found to be infected before symptoms appear, must be slaughtered to prevent the disease spreading. Farmers are deeply attached to their animals and having to slaughter a herd can be even more distressing emotionally than the financial loss. The disease is most readily spread from cow to cow within the tight confines of cattle sheds.[2] It is also spread when slurry from infected herds is used to fertilize pastures, which is how badgers and other species become infected. We know that bTB and other similar diseases spread rapidly when cattle are kept indoors on industrial-scale dairy farms. Furthermore, calves are removed from mothers within days of their birth, adding to the stress already caused by overcrowding. Stress could weaken both cows' resistance to infection and the effectiveness of diagnostic tests for bTB.

---

2 Intensively farmed dairy herds kept crammed together indoors suffer far more bTB than beef herds kept mainly outside.

We saw during the COVID-19 pandemic the importance of diagnostic testing, quarantining infected people from those vulnerable to the disease, and rapidly developing effective vaccines. So it follows that monitoring of cattle for hidden bTB infection is essential, yet the current skin test is only successful in detecting bTB in around half of infected animals. Much more sensitive tests have been developed, but Defra's current regulations largely prohibit their routine use. Since farmers are financially compensated by the government for every cow slaughtered due to bTB infection, these outdated policies presumably remain in place so that the government can avoid paying out more compensation. Ensuring that uninfected cattle are protected from infected ones, and putting every effort into development of a cattle vaccine, would be highly effective strategies to limit the spread of the disease. Despite this, efforts to control bTB have centred on culling badgers rather than targeting cattle-to-cattle transmission.

Badgers have been persecuted for centuries, by badger-baiters who have made them the centre of a gruesome spectacle, and by fox-hunters for whom badger setts act as hiding places for foxes. The presence of badger setts on farmland may interfere with land management, but farmers are not permitted to plough them over because of the 1992 Protection of Badgers Act. It is certainly more convenient and requires less financial commitment to cull badgers rather than change traditional agricultural practices.

A major government study carried out between 1998 and 2005, the Randomized Badger Culling Trial, compared rates of bTB in culled and unculled zones in the south-west of England, where bTB prevalence is especially high. At the end of the trial, an independent scientific group overseeing the study concluded that culling badgers would have no effect on the disease and should not proceed. The Conservative government decided to ignore expert scientific advice and in 2013 and 2014 performed pilot studies of culling in Gloucestershire and Somerset, the results of which again showed little reason to proceed.

Nevertheless, Defra rolled out badger-culling on a massive scale and more than 250,000 badgers, about half the estimated UK population, have now been killed, at a cost to taxpayers of around £100 million and an unimaginable cost to ecological balance. Subsequent work has shown no difference in bTB incidence or prevalence in culled versus unculled areas. Furthermore, bTB is better controlled in Scotland and Wales, where they concentrate on preventative measures in cattle and don't cull badgers.[3]

Badgers are important. They help to control rodents, small mammals and insect populations, and by digging tunnels help to improve soil drainage, aeration and plant growth. They are fundamental to a balanced ecosystem and are a much-loved native species. I simply had to challenge what I saw as a senseless and unscientific policy. So I became part of a Badger Army working hard to protect badgers.

I started with Freedom of Information requests to establish what evidence – if any – Defra was using to support badger-culling policies. This information was never provided, even after my challenges to Defra reached the First-tier Tribunal level.[4] Feeling

---

3 According to UK government data, cattle herd TB incidence in England fell from 9.8 per cent to 7.3 per cent between 2012 and 2023 (a fall of 25.5 per cent). In Wales, *without culling badgers*, the figures were 10.0 per cent in 2012, 6.8 per cent in 2023 (a drop of 31.3 per cent).

4 In UK Freedom of Information proceedings, a First-tier Tribunal is the next step after the Information Commissioner's decision. It is held in front of a judge, which can be either in a Tribunal Court or in a video hearing. The next and final stage is the High Court, where it is mandatory to be represented by a King's Counsel (KC), at costs the petitioner can rarely afford. Hence the government usually wins, at the expense of taxpayers. A summary of our arguments and an account of our legal challenges to Defra can be found in a YouTube presentation 'Bovine TB and the Badger Cull Update December 2022'.

that science-based arguments had fallen on deaf ears, I reasoned that the only way I could realistically protect badgers was by using nonviolent direct action. This has seen this immunologist patrolling badger habitats late at night, working with others to prevent killing.

We have covered many miles of countryside with hunt saboteur groups looking for and removing bait points and cage traps. Our expeditions always run the risk of opposition from pro-cull farmers, especially when they involve trespassing on private land. The worst moment? Coming across the gruesome sight of a dug-out badger sett used for badger-baiting. The best moment? As dusk was falling at the end of another exhausting day, we discovered a cage trap which had caught an adult badger. Releasing that badger, which would have been shot the following morning after a frightening night in captivity, felt powerful and rewarding.

***

After decades working to combat infectious diseases, knowing that effective, collaborative action is both possible and needed, but that our systems and governments are seriously letting us down, I find myself fearful of what the future holds. There have been many spectacular advances in the treatment of chronic diseases in recent times, but epidemics of acute infections are bringing new challenges as novel species of viruses, bacteria and parasites emerge from a challenged environment. As the climate warms, the ranges of disease-carrying species, such as mosquitoes, expand, so that tropical infectious diseases like malaria and dengue fever are spreading increasingly widely. Melting of the permafrost could unleash microbes never previously encountered by humans, with potentially devastating consequences. Humans are encroaching on and destroying wilderness areas in search of more land for development and for crops and livestock. We are not only losing the trees that sequester greenhouse gases but are also coming into closer contact with previously unknown pathogens harboured by indigenous animal species. Have we learned any lessons from the origin of the AIDS-causing virus HIV, which jumped into humans from macaques sought for bushmeat in the sub-Saharan African wilderness? Or from the SARS pandemic that had its origins in the wet markets in southern China, passed to humans from civet cats being sold for food?

We should be very concerned about the collective government failure to appreciate the danger of further epidemic infectious diseases as the world becomes hotter, more stressful, and more damaging to physical and mental health, and as humans encroach further on nature. We have the scientific knowledge to help control new infectious diseases, and decisions on implementing protective, preventative and therapeutic measures must not remain solely in the hands of politicians. Mistakes made during the COVID-19 pandemic have clearly shown that science must be allowed to drive climate and health policy; politicians will have to work closely with expert scientists to stem and reverse the destruction of the natural environment and improve the chances of human survival in our threatened world. Politicians are not experts and can be vulnerable to the vested interests of powerful lobby groups such as the pharmaceutical industry in the case of infectious diseases, the fossil fuel industries in the case of the climate crisis, and the National Farmers' Union in the case of badger-culling. To be heard above the noise of these forces – and to serve the interest of humans and animals alike – scientists must raise their voices higher.

THEY KNOW

Jen Murphy, PhD

Jen learned about the greenhouse effect not long after receiving her first chemistry set, aged seven. Convinced that governments would be doing the work needed, she studied for a degree in chemistry, worked as an accountant, raised a young family and completed a PhD, publishing in the field of population data science. It took her thirty years and one IPCC report to realize the scale of the crisis, and she first stepped out into the road as a climate activist with Extinction Rebellion in 2019. Jen now works as a science teacher in a UK state secondary school and takes great pleasure in being in charge of the fun fizzy bangy stuff.

It's a wonderful place, a school. There's nowhere quite like it for energy. There's always a buzz in the corridors, never-ending excitement in the break queue, and a professional and intellectual challenge that changes with every lesson. The atmosphere is febrile and curious; students malleable, constantly growing, adapting and changing. It's a privilege to go to school. Especially when you're the teacher.

The thing is, all of us teachers, we also went to school, and when I was at school I learned something really truly terrifying.

Our teacher told us one afternoon as we sat down, pencils at the ready, there were these things called greenhouse gases. They were coming from fossil fuels, and they were causing the climate to change. Our planet was warming up. The greenhouse effect was going to mean no more polar bears. And polar bears were cute.

Bad news all round.

Even worse, there were also these things called CFCs[1] and they were busting a hole in the ozone layer, which meant that the people of Australia were literally frying when they went outside. Oh yes, and the Amazon rainforest was being felled at a rate of a country the size of Wales every three minutes by illegal loggers and those naughty folk from McDonalds for all their burgers. (I exaggerate, but you get the picture.) The oil was running out. The gas too. Mrs Thatcher had shut the mines. We needed to cut back on hairspray, and stop buying mahogany furniture and fast, otherwise we'd be in big trouble.

I was worried. Really worried. The planet was in meltdown and we were running out of energy! We made posters about

1 Chlorofluorocarbons (CFCs) are chemical compounds of carbon, chlorine, fluorine and hydrogen and were widely used as aerosol propellants and refrigerants. They are classed as 'ozone depleting substances'. The Montreal Protocol, ratified in 1987, is a UN treaty for the discontinuation of their use. As these chemicals were phased out, others which do not deplete the ozone layer so much were developed such as HFCs – hydrofluorocarbons (like a CFC but without the chlorine). Unfortunately, these are also extremely potent greenhouse gases, and so amendments have been needed to the original Montreal Protocol to also phase out the use of these chemicals. Note that popular brands of non-stick cookware use CFC replacements, but when heated to a high temperature they bond with carbon in the food being cooked to release, yep, you guessed it, CFCs. So your burnt dinner isn't just a culinary disaster, it's also cooking up an ozone-depleting global warming storm.

banning CFCs. Drew pictures of the hole in the ozone layer. I'd like to say I vowed to never eat a burger again but that would be a lie. A trip to a burger joint was a big treat and anyway, no one seemed to think we needed to stop that kind of thing and the good people on the science telly programs said that any moment now we'd have nuclear fusion, and jet packs and solar-powered aeroplanes. No need to panic. Science will save us. Back to my chemistry set on the kitchen table.

It was bad news, back then. My 1990 *Blue Peter Annual* told me so. 'There's no doubt about the causes of the climatic changes,' I read. 'It's up to us to do what we can to stop them.' But no one seemed to be too worried. The grown-ups kept on as usual – and by the nineties we hardly talked about it in our science lessons at school. It was the 'and finally' section of the news. Not the main event.

So here's the science bit. We need the greenhouse effect – it is what makes our planet habitable. Without it life as we know it would not exist.

When radiation from the Sun reaches our planet, some is reflected by the atmosphere and the surface of the Earth, and some is absorbed by the Earth's surface, making it warmer. Hotter objects emit shorter length radiation than cooler objects – in the case of the Sun we have visible light and near-infrared (this is radiation you might know as 'heat' rather than light, but in reality it is all the same thing; it is just that human eyes have evolved to use visible light for sight). It is why to us it looks like the Sun shines but the Earth and the Moon[2] do not.

When the Earth absorbs this radiation, it re-emits it as longer wavelengths than originally were absorbed (because it is cooler than the Sun), in the far infrared part of the electromagnetic spectrum. Molecules of gas in the atmosphere absorb this longer

2 Remember, moonshine is reflected sunlight. And a fairly stiff drink.

wavelength radiation and re-emit it in all directions, trapping the heat within the planet's system. There are lots of molecules that can do this – not just the chemical nasties like CFCs that you might first think of – water vapour is a greenhouse gas, and carbon dioxide, too.

Some people think the greenhouse effect is the same as pollution, but even without humans there is carbon dioxide in the atmosphere – it is part of the natural carbon cycle of the planet. The carbon cycle is complex but the main point is that in nature there is a balance – plants and vegetative matter take in carbon dioxide and use it to grow, harnessing sunlight in a process called photosynthesis. These plants form the bottom of the food chain. When an organism dies, it decomposes back to its constituent elements – some of the carbon might be kept in the soil for a while but eventually it is returned to the atmosphere as carbon dioxide, to start the cycle again. In the words of Disney's Mufasa in *The Lion King*, 'When we die, our bodies become the grass and the antelope eat the grass.' (Carbon dioxide also cycles in and out of the sea, but we can leave that for another time.) When plants and animals died millions of years ago, in certain geological circumstances their carbon was buried, compressed and then stored as coal, oil and gas. This carbon store makes up the fossil fuels we discovered in the industrial revolution and which power our track to climate destruction in the present day. Carbon sequestered in the ground over millions of years as fossil fuels has been dug up and released into the atmosphere in a matter of decades by human activity. This rapid rise in the concentration of atmospheric greenhouse gas means that more and more of that longer wavelength infrared radiation is staying trapped inside our atmosphere. With all that extra heat, the average temperature of the planet is increasing. This means that we see disrupted weather patterns, the rise in sea levels and more – all of the consequences we're getting used to hearing on the news.

Back to the present. I stand in front of my Year 7 class. I am nervous.

I know that there are some people, and many parents, who don't like it when you talk climate. 'It's too political. It's not for children. It's being over-dramatized. It's not true.' I'm the token greenie at most gatherings. I am the person people apologize to before flying to New York for the weekend or ordering a huge steak.[3]

The students who see an environmental protest in the news and think they might like to add their voice are few and far between. Even the straight science version of our climate emergency is intrinsically political. How was I going to deliver the bad news and stay inside teachers' standards? How could I handle this now that we know what I learned back in the 1980s might happen in the future has actually come to fruition?

'We're going to learn today about something very important – it's about how our use of energy resources is causing climate change.'

Josh[4] puts his head on the desk.

'Head up, Josh,' I say. 'We're going to listen and then do some activities.'

'But I can't, I can't,' he says.

I leave him resting, while I get on with a starter to find out what the students already know and think. In a quiet moment I whisper to him, 'Are you OK? Do you think you can lift your head up and get started?' He is clearly distressed. His neurotype sometimes means that school is difficult and I don't know what might have happened in the hours before this lesson.

---

3 Please stop apologizing to me. Your kids will inherit this Earth. If you have to apologize (and I really think you shouldn't – you didn't cause the climate emergency, you are also a victim here) – say it to them.

4 Not his real name.

He lets out a wail.

'I hate it and hate it and hate it,' he shouts, louder now. His bench mates are looking over. 'I thought it was supposed to be getting better and it's not. We are always learning about climate change. Nothing ever changes. Nothing ever gets better. What are we going to do?'

I have nothing for him.

I know that feeling. The anguish. The grief. The helplessness. Josh feels the climate crisis like we should all feel the climate crisis. Raw. Honest. Terrifying.

'I'm sorry,' I say. 'You're right and I'm worried too.' But nothing I say can ever be enough.

My Josh-moment happened in 2019. A fairly jolly jaunt into an Instagram challenge to go plastic-free led me down a path of enquiry and self-reflection that landed me, on a summer's day with my own children looking on, slightly embarrassed as I sobbed on the floor of the Greenpeace tent at a festival. I felt just like Josh. I just kept saying, 'I thought the grown-ups had it under control.' I really did think this – but I've read the 2018 IPCC report and there is one thing that is abundantly clear: the grown-ups did not have it under control. The grown-ups still do not have it under control.

Until that point, nothing I had done mattered. Every petition I had signed. Every letter I had written. Every separated recycling bin. Every second-hand purchase. Every careful repair. Even those 'Ban CFCs' posters displayed in the classroom. Not one of these things had changed our course. I am a grown-up now. I absolutely do not have it under control. I didn't know that what I was doing wasn't working – with all my scientific training, my interest in the world, my 'green' credentials, I simply didn't realize just how bad it was.

The kids, though – they know. We spend the rest of the lesson gathering key words – they know them all. Their knowledge is impressive and humbling. Their concern is well founded, carefully articulated. We calculate our personal carbon footprints (it is an affluent school but there is still a huge gap between the lower and higher income families).

They understand about deforestation. They have fun playing a carbon balance game with balls. They know about mass migration, crop failure, flooding, starvation. Some of them have family in more affected places in the Global South. They have seen the dust of a drought-ridden field, felt the fierce unbearable heat of a summer in one of the hottest years on record.

They all know, but when I ask them to make suggestions for change, I am floored.

'I could go and play Xbox at someone else's house to save my electricity.'

'I could turn a light off.'

They're a start – share things, save energy. I probe a little deeper, 'Could you maybe choose the vegetarian option for your school dinner?' Their faces aghast – no ham on the school pizza would apparently be unforgivable. 'What about walking to school instead of getting a lift?' Met with derision. 'You could write to your MP?' Blank faces.

They know the facts, but they are so robbed of agency and so helpless in the face of such overwhelming bad news, the best they can do is just carry on. In one study, 75 per cent of young people reported feeling frightened about the future, with 83 per cent stating that they did not believe people had looked after the planet. These children and young people feel betrayed by government and that they themselves have no power to affect change.[5]

Josh's expression becomes more and more pained. I make a

---

5 See Hickman, C. *et al.* (2021) 'Climate anxiety in children and young people and their beliefs about government responses to climate change: a global survey.' *The Lancet Planetary Health.*

mental note to let another staff member know how distressed he has been, that this might impact his learning for the rest of the day, even the week.

But in this moment I can't help thinking, they're right and actually, so what? As an individual it is really hard to have any influence on the powers that drive this crisis, and it certainly isn't the responsibility of the twenty-eight eleven-year-olds I have in front of me. They did not cause this crisis. If the rest of today's learning is disrupted, is there really any more important lesson than the one Josh has just learned – that the first step to facing this, is to feel it? Is our whole school system just churning out more consumers, more people skilled in jobs we don't need doing? Is this semi-education leaving them with all of the despair, and none of the agency? Does this mean they feel so helpless in the face of the challenge that they have only two choices – endless unproductive despair, or the relative emotional comfort of quasi-denial? Should I be teaching them about the enhanced greenhouse effect, or would this time really be better used and their long-term well-being better served by learning about social theories of change, showing a more empowering and hopeful version of what the future might hold?

When I got up from the floor of that Greenpeace tent, I realized that it was time to stand up and be counted. To put myself in the way of the climate-wrecking machine. Manchester. London. Warrington. London again. I sat in the road. I sang in a police kettle. I occupied a museum. I shut down an insurance company. I disrupted a big bank's AGM. I walked out in front of the London traffic with just a sign and hope to protect me from the wrath of drivers. We made the news. The *Evening Standard*. The *Guardian*. The *Financial Times*. The *Daily Mail*. People talked about it. I had to be brave, but I was surrounded by people who get it. Others who've had their Josh-moment somewhere down the line. It feels good to be together, a community – I make new friends every time I act, even if just for a few hours. There is love and acceptance – we're all learning. The climate crisis is not our fault.

The teachers' standards state that I have a duty to safeguard children's well-being. When it comes to the biggest crisis we all face – a warming climate – governments of all persuasions have failed to do this. I stand up for so many reasons: it's my professional duty, it works, I want to show others that we can, and in those tense moments of action, I see the glimmer of that empowering and hopeful version of the future.

We ran out of time to explore any further ideas in that lesson so they chose the pledge that is perhaps the easiest and yet most difficult of them all – to go home and talk about what they had learned. To say how it made them feel. To talk about what they might do. A tiny step in the direction of action.

I HAD A CHILD IN THE PLANETARY EMERGENCY – NOW WHAT?

Viola Ross-Smith, PhD

Having always loved the natural world, Viola followed the academic route through university, but decided that academia itself wasn't for her. She worked as a seabird ecologist for a UK environmental NGO for some years. However, she felt that communicating through peer-reviewed papers and the commercially confidential reports that are part of 'grey literature' was not her forte. Nor did it help her to make a difference. She gravitated towards science communications and has been working full time in this field since 2015.

One scientist talks to three older scientists, myself included, about the decision to have kids:

'I'm in a tight spot. I've always liked kids, but my partner isn't keen. It feels like I have to make a choice, and hearing your experiences could help me with that.'

I cast my mind back a decade to when I was grappling with those same thoughts. I had dithered about having a baby. One by one, just about everybody my age seemed to be going for it, and I stalled, trying to convince myself that I'd be better off childfree.

My concerns about the climate and nature emergencies were only part of that. I was also grappling with more day-to-day reservations. Would I be a good mum? Did I have enough of a support network to make it work? Could I afford it?

Another of the scientists chimes in: 'I got pregnant in 1999, knowing full well that the climate crisis was in motion and that just by living in the UK you take up more resources than the average person in the world. So having a baby adds to that. My animal self is an important part of me. I learn from it all the time, especially in terms of empathy for all creatures, so I find acceptance of its motivations quite natural. I never questioned my wish to have kids, even though I had friends who did. Rationally, I totally get why people don't have kids.'

My 'animal self' – that's it! That's what I had tried, and failed, to suppress. Maybe if I'd also been able to accept it, I would have saved myself a lot of agonizing. Of course, reproduction has been programmed into us all through billions of years of evolution, and I shouldn't chastise myself for not being able to withstand it.

In my case, I had the feeling of being hijacked by my own ovaries. I fell in love with somebody who already had two young children from a previous relationship. His children were (and still are) brilliant, but it felt painful to be doing the 'mum thing' while trying to keep a respectful distance because I wasn't their mother.

And I was constantly being mistaken for her. I was looking after her children, doing the work, making the compromises, and something was missing. I started mourning another unborn child every month when my period came. The overheard words 'Viola isn't very good with children because she doesn't have her own' bounced around in my brain and needled my heart whenever I was with my step-kids.

My partner knew how distressed I was, no matter how much I tried to rationalize it and no matter that I could see all the perfectly valid reasons why not bringing another Western European into the world was a good idea. We tried for a baby.

Unlike people who are childfree not-by-choice, we were very lucky: it worked. And my child is everything to me. I feel privileged to be his mum and he makes me so happy. But when I look into his beautiful brown eyes, especially when he's seeking reassurance, I also feel guilty and heartbroken for even bringing him into the world, given the catastrophes he could be enduring by the time he's my age. Assuming he and his generation make it that far.

I'm an ecologist. I remember finding out about 'global warming' (as it was called in those days) and ecological destruction when I was a child in the 1980s. I felt galvanized by news footage of habitat destruction in the Amazon and seas red with the blood of whales and dolphins. I took a Greenpeace collection tin to friends' houses, drew pictures of rainforest animals for my classroom wall, and hoped the grown-ups would sort it out. This was the era of the Montreal Protocol on Substances that Deplete the Ozone Layer and the moratorium on commercial whaling, after all.

I could see things weren't changing fast enough as I grew older, but I also had a sense that the 'real' problems were a long way off and there was still time to turn things around. Right up until 2015, I still felt optimistic following the Paris Agreement, when 195 countries signed a United Nations treaty to limit the global temperature increase to less than 2°C (1.5°C if possible) compared to pre-industrial levels.

However, in the years since, my understanding has shifted dramatically. The forest fires, the floods, the heat domes, the extinctions – all the horrors I thought were farther in the future and hoped were avoidable were increasingly happening here and now, and all at once. Disaster upon disaster.

And then in 2019, Extinction Rebellion burst onto the scene. I started finding footage online of people who looked and sounded like me but were taking action in a way I'd never imagined that I could – members of Scientists for Extinction Rebellion. I reached out to them and started to get involved. But by then, I had already become a mum and more questions were swirling around in my mind.

How do I bring my brilliant boy up so he understands what's already happening, let alone prepare him for what's coming? How do I equip him for the future without filling him with despair? I'm not sure, but I feel like I'd be doing him a huge disservice if I wrapped him up in cotton wool and pretended none of it was happening, so I'm not.

When he asks me questions, I give him honest answers. I don't want to scare him, but I also don't want to lie. It's actually not all that hard to explain climate change and biodiversity loss to him – he's witnessing it with his own eyes already. The days in summer 2022 when his school closed at 1 p.m. because the heat was too extreme for learning, and the 'once in every fifty years' flooding events in our local area, that now seem to be happening nearly every winter. The springs when he accompanies his dad and I as we monitor bird nest boxes, and he can see the effects of unseasonal weather and lack of insect prey when he lifts the lid of box after box to find the bodies of stunted and under-developed chicks. This is normal for him.

I have also brought him to family-friendly Scientists for Extinction Rebellion actions, hoping it will help him as he grows up to see that lots of people are putting pressure on governments and corporations to change. Maybe that will help him to not lose hope.

My son inspires mine. When I first got involved with Scientists for Extinction Rebellion, he was in the forefront of my mind. I am scared of being arrested or getting known as a troublemaker, because I need to be able to care for him. I can't go to prison, and I can't afford to lose my job. But can I afford not to act when his future is at stake? What kind of parent would I be if I didn't try everything in my power to change things for the better? So, I have found myself taking more risks and sticking my head above the parapet. I attend protests, speak in front of crowds at rallies, and push my workplace to shift its stance on the planetary emergency and update its policies. This is not in my nature, but I feel compelled to do it.

My choice to have a child was an act of hope. My choice to act with Scientists for Extinction Rebellion is, too. I still hope we can turn things around. I still hope my beloved child won't experience ecological and societal collapse and all the horrors that would unleash. Maybe by bringing him up to see what's happening, he can be part of the solution? If only it were that simple. But I need to keep on hoping, so I can keep going for as long as I have the power to do so and keep on pushing for the changes necessary for a better world. So, I can look into my son's beautiful brown eyes as the realization of what's at stake continues to dawn on him, and tell him honestly:

'I did my best. I tried.'

WISHFUL THINKING? HOW TO HAVE HOPE IN A PLANETARY CRISIS

Neal Haddaway, PhD

Neal is an environmental photographer, photojournalist, and researcher. After a twenty-year career in environmental science, he turned to photography as a medium for working towards societal change. His photography and journalism explore the role contemporary human society plays in the destruction of nature, and the emotional toll caused by a scientific awareness of the impending planetary crises. His research examines the potential for bias and misinformation in agricultural photography in mainstream media.

Do I feel hope about the future? It's a question I sometimes get asked when I tell people I work in environmental science – they're often looking for reassurance from an expert. It's also a question I ask myself. Mostly I try to change the subject subtly, to divert my attention. Much like feeling for a broken tooth with my tongue, though, it's hard not to return to it. But what does it mean to feel hope? Is it OK to feel hopeful? Am I wrong if I don't? How do we carry on if there's no hope?

In 2021 I was conducting environmental research, working in what we call the science–policy interface – trying to integrate rigorous scientific evidence into how we mitigate and adapt to climate change. I became disillusioned with my work; I no longer felt that what I was doing was worthwhile, or worthy of the public funds going to pay my salary. If policymakers couldn't even follow the simplest and most obvious of science – that we cannot begin new fossil fuel extraction projects – how could we claim to be designing a better system to get science into policy?

This feeling of hopelessness crept over me slowly, starting as a niggling doubt, culminating in the decision to leave academia and its accompanying experiences of trivial metrics, toxic politics and ineffectiveness. I felt a huge sense of loss – environmental science and conservation was all I'd ever wanted to do, and all I'd ever known. I wanted to carry on – I felt a deep sense of duty to protect the natural world and make up for centuries of my ancestors' wrongs. But I lacked the drive to maintain the excessive levels of work demanded by my academic institutions.

So, I applied to study for a master's in photography – a serious hobby for much of my life and something that I thought might act as therapy and a creative outlet for my grief. As part of my application, I needed to develop a photography project. Wracking my brains for a meaningful and engaging topic, I was explaining my choice to leave academia to a colleague when he shared his own feelings of fear and hopelessness. Each time we spoke, we discussed our despair and deep sense of loss, but we also grew closer – the conversations uncovered pain, but left catharsis in their wake.

I realized that so few of us, as environmental researchers, speak about our climate emotions – the range of feelings we experience as our knowledge and awareness of the state of the planetary crises grows. In my project 'Hope? and how to grieve for the planet' I interviewed twenty-five environmental scientists and communicators, exploring common themes of guilt, anger, sadness and frustration. Each person discussed three emotions they experienced, I photographed them as they spoke for several hours, and we collaboratively selected three portraits, one for each word. Several people cried, including me – it was a challenging and humbling experience. Most participants said the process was cathartic, although several felt somewhat shell-shocked for a few days afterwards.

During the course of my project, someone on social media told me about Joe Duggan's project, 'Is This How You Feel?' A science communicator, he had asked more than seventy climate scientists to compose handwritten letters describing how they felt about the climate crisis. The letters make for an incredibly moving and sobering read. One thing that I saw across Joe's project and mine was that many people reacted negatively when talking about whether they felt hope. Several felt hope was dangerous or offensive in some way. I found this particularly interesting – if we, as environmental scientists, don't have hope, then how can we stay motivated to keep working in such a difficult field?

The question piqued Joe's interest, too – I reached out to him in 2021, and with a colleague, Nic Badullavic, we set about analysing the letters from the 'Is This How You Feel?' project. We used thematic analysis to code emotional statements and references to hope. We found that, unsurprisingly, negative emotions were far more common than positive ones. Interestingly, though, we found frequent references to hope. Could this be evidence that climate scientists were optimistic about the future? We took a deep dive into the references to hope, and found that it was being used in two different forms – what we called 'wishful' and 'logic-based' hope. More often, people referred to *hoping that things would get better*, and wishing that things would improve. Less often would people express logic-based hope and implicitly or explicitly say that *something gave them hope*. We also noticed that the few dozen scientists who wrote a second letter for the project four years later expressed much less logic-based hope, although their wishful hope remained the same. It seems perhaps that as time went on, we had less to base our hope on.

I sometimes feel hope for the future. When I photograph the enormous lignite mines near Cologne or the vast greenhouses in Almería I recognize what philosophers and art scholars refer to as the 'industrial sublime' – a deep feeling of awe at the scale of human ingenuity (and destruction). Although viscerally painful for me, seeing what humans can do to the planet does give me hope – evidence of our ability to drastically modify the landscape on an unimaginable scale. If used for good, this power could create a very different future than the end of civilization that neoliberalism and corporate greed are sprinting towards.

Hope is a tricky beast – sometimes an uncertain, vague faith and other times an evidenced, concrete motivation. Cultures are built on the foundations of collective beliefs in unseeable and unmeasurable concepts and values, but they're also based on powerful stories of hope. As environmental scientists, we know what needs to be done to avert a climate catastrophe and planetary collapse. And yet we must observe the continued inaction and destructive decision-making of our leaders, and the more frequent and severe negative effects of the worsening climate crisis. There are perhaps no grounds for hope, and yet hope is exactly what we need.

THIS CHALLENGE NEEDS A NEW APPROACH

Jeff Waage, PhD

Jeff is a Professor of International Development at the University of London. He started his scientific career as a Lecturer in Biology at Imperial College in 1978 and has since worked in agriculture and development for UK universities and with the Commonwealth Agricultural Bureaux. He has been on scientific advisory groups for the Department for Environment, Food and Rural Affairs (Defra) and Natural England, and has spent much of his career providing evidence for international policy to the United Nations (UN) and World Bank. He is now largely retired and works on biodiversity restoration projects in London and with Scientists for Extinction Rebellion.

I came to climate activism after a scientific career of about fifty years spent working mostly in the Global South on agricultural development, environmental conservation, and related areas. Through a sequence of roles in universities and international organizations, I found myself increasingly drawn into providing scientific evidence to policymakers. By the time I stepped back from my last academic job, I had served on scientific advisory bodies of UK government departments and UN programmes.

I decided to 'retire' around the same time that climate change became front-page news with the Paris Climate Summit in 2015. I was an ecologist and had not been keeping up with climate science. As I learned more about it, I realized that achieving climate change commitments would involve a science-to-policy process much bigger than any I had ever worked on. Compared to my experiences with feeding science into national and international policies on issues such as biodiversity, pesticide use or nutrition, climate change had a scientific evidence base that was much more complex. Much of the evidence came from mathematical modelling of future scenarios rather than clear experimental results, a particular challenge for policymakers.

Over the course of my career, I had become slowly disillusioned with the process by which science contributes to government policies at the national and international level. It is a slow process which depends on evidence reaching those politicians who have both the inclination and the time left in their tenure to do something. My experience with challenging pesticide misuse taught me that lobbying by companies with vested interests could really slow down policy uptake by governments. I felt that the urgency of the climate and ecological emergency demanded something else, particularly given the power of the fossil fuel industry to interfere.

In November 2016, I had a transformative experience. I was invited to a Rockefeller Foundation workshop to advise them on what environmental research they should be supporting to best inform future policy. We sat there for a couple of days, brainstorming and feeling very important, as scientists do when they are enjoying posh surroundings and 'saving the world'. But on the third day, we woke up to the news that Donald Trump had just won the US presidential election. Over breakfast, we agreed to change our plans and spend our remaining time exploring the role of science and scientists in a world where the science-to-policy process was broken. Out of that came an idea that we, as scientists, should be focusing more on bringing science to citizens and communities, and not just to governments and politicians. I had never done that, and it sounded challenging and new. I came home with an idea for my post-career career.

I started volunteering locally, going back to my lifelong love of nature, contributing to biodiversity projects in urban green spaces. It was an easy and logical step in 2018 to join Extinction Rebellion, because of its focus on engaging citizens and demanding that governments listen to the science.

I now work with a community of scientists in Scientists for Extinction Rebellion that motivates me and helps me to become a more effective scientist–activist. My colleagues are extremely bright, highly motivated and staggeringly creative. Most of them are holding down day jobs and are working with XR at some risk to their careers, quite unlike me. And they are much younger! I would say there is a good thirty to fifty years between me and most of them.

My age and career leave me with a particular view on where this is all heading and how I want to contribute. In the various science-to-policy processes in which I have participated over recent decades, I have come to know people working on all sides of these issues – in harmful industries, ineffective governments and dynamic campaign groups. I have never found bad people, just bad ideas and institutions. So I have come to the view that addressing the climate and ecological emergency is not a battle of good and evil. It is about changing views, and it requires strategy and engagement, not simply opposition. For me this means putting myself in the shoes of the people I am trying to engage with or change, and trying hard not to slip into anger and frustration with them, as that makes me less effective. I see some of my younger colleagues living with an intense sense of urgency, and suffering from anxiety and exhaustion, and I want to help them to slow down, and to pace themselves for the long struggle ahead, with its many small, but accumulating wins. I like the XR slogan 'Love and Rage' as it captures for me the need to balance urgency with peaceful energy.

Personally, I don't think that we will address climate change effectively within a safe timescale. Ideas, institutions and people just change too slowly. But my experience tells me that we will win eventually if we maintain our steady effort. I have to accept that the world which we will finally win for humanity and nature will be more degraded than our present one. And I have to accept that this loss happened 'on my watch' as an ecologist these past fifty years, because I did not take serious action – until now. My children will never know all the natural beauty of our planet and its people that I have seen. They have grown up in a world where much is already gone. But that is my felt loss, not their lived experience. Acknowledging all this, and letting go of my own sense of loss helps me look forward with determination rather than with anxiety. If I can help to ensure that my children or grandchildren have some nurturing nature to make their lives more feasible and rewarding, that will be a worthwhile win.

PART 2

SCIENTISTS MAKING LIFESTYLE AND MINDSET TRANSITIONS

'WHY I NO LONGER FLY'

FROM TRIPE TO TOFU

Caroline Vincent, PhD

Caroline is a dual French and British citizen. She was educated in France, with a degree in agricultural engineering and a PhD in biology from Pierre-et-Marie Curie University in Paris. She worked as a postdoctoral researcher at Cambridge University, UK, for three years before obtaining an MBA from INSEAD Business School in France. After returning to the UK, Caroline had several managerial positions and, for seventeen years, she worked as an independent consultant building forecasting models for the pharmaceutical industry. In 2018, Caroline joined Extinction Rebellion and two years later she quit her job to become an activist. Caroline has participated in numerous actions with Scientists for Extinction Rebellion.

I love food. In my family it was an obsession at the very centre of our lives. I was brought up in the Loire Valley in France, where we were never far from the countryside. From an early age I learned that each season brings new culinary delights. Delicious fat stems of white asparagus at Easter, ripe juicy greengages with sun-cracked skin at the end of the summer holiday, and roasted chestnuts too hot to hold in the winter. The start of autumn was particularly exciting, with long walks in the forest searching for wild mushrooms and the added bonus of a fleeting glimpse of a deer or wild boar. Fruits, vegetables and fungi all had their seasonal place in French cooking, but the main event was meat, fish and dairy – and a lot of it! Going to the local market is one of my fondest childhood memories. Before I was tall enough to see over the fishmonger's counter, I was helping my rather bossy grandmother to pick out the biggest periwinkles from the enormous mountain of shiny black shells that glistened like precious stones. Then there was the butcher, a giant of a man with a stern look wielding terrifying sharp knives. He knew exactly where each cut of meat was coming from, right down to the breed and the farm that raised it. He delighted in gently explaining the perfect way to cook that particular cut of meat while ferociously carving it. But my real favourite was the charcuterie, a temple to all things pig, with every part of this animal on display in one form or another, from a simple boiled ham to the most elaborate pie. By the age of five, there was not a body part of the pig that I hadn't consumed with relish, including the trotters and ears. And yet, this little girl dreaming of becoming a vet also had a profound love of wildlife and was never happier than when out in nature, looking for frogspawn or climbing trees to look for birds' nests. In her mind, all the delights of carnivorous gastronomy co-existed quite comfortably with an unconditional love for animals. And so it went on without any qualms for several decades, long after the little girl had become a woman.

After finishing my PhD in Paris, I was awarded a grant to do research in Cambridge. I packed my bags into a battered old Renault 4 and crossed the Channel, expecting culinary torture in the land of Marmite and baked beans. I swore to all my French friends that I would be back after twelve months. That was thirty-five years ago and I am still living here in the UK. What went wrong? Well, I fell in love with a country where the people are kind, open-minded and nicely eccentric. I also fell in love with a British boy, with whom it turned out I would share the rest of my life. When I took him to France to meet my parents for the first time, he was initiated into the delights of foie gras, steak tartare, and tripe, and he was hooked. Back in Marmite land, I saw over the nineties and noughties how celebrity chefs like Delia Smith and Jamie Oliver transformed British culinary culture. You only have to watch *MasterChef* and *The Great British Bake Off* to realize the UK has come a long way from the days when a prawn cocktail was the height of gastronomic sophistication. As I was witnessing this transformation, I was also starting my own culinary journey of discovery. It was around that time that I first encountered vegetarians and vegans, very rare and endangered species in France! One evening I cooked for a group of friends. 'Best beef bourguignon I ever ate,' said one of them. 'Oh, thank you, that is quite a compliment,' I replied. 'Not really,' he continued. 'I'm vegan and this is the only bourguignon that I've ever eaten!' Veganism and vegetarianism were so off my radar at that time that I hadn't thought to ask our guests if they didn't eat meat. My culinary awakening was about to begin …

It all started with a newspaper article on climate change. Of course, climate change was something that I was already aware of but it was at the back of my mind; surely the politicians running the world had it 'under control'. After all, in 2015 in Paris, most countries had promised to take action in order to limit the temperature increase to a safe level. All was good, or so I thought! Then in 2018 I read about a newly published report on the impacts of global heating. It opened my eyes to the devastating consequences of a 1.5°C rise above preindustrial level and, worse still, even a few tenths of a degree above this.

And we were on course to reach far beyond 1.5°C, unleashing a cascade of irreversible events including a rising sea level, the destruction of coral reefs and the transformation of rainforests into savannas. From then on, I read avidly about climate change and, within a few weeks, had psyched myself up to climb power station chimneys with Greenpeace to protest. But Greenpeace only offered me actions like picking up litter or unwrapping plastics outside supermarkets. In my opinion this level of engagement was far from sufficient to address the emergency we were facing. I felt that society needed an electric shock to change its course. To do this 'we were going to need a bigger boat'. This was when I joined Extinction Rebellion. But even then, I did not question my carnivorous lifestyle. Probably because animal farming and eating meat were so culturally ingrained that I had never questioned them. Not even during my agricultural degree in France when I worked on farms producing meat and saw first hand what goes on in a slaughterhouse.

As an adult in England, my household was lucky enough to be able to afford meat from selected organic farms, where I reasoned at the time that most of the animals had had a decent life. Our bacon had recently been rolling around in a muddy field and our Sunday roast had spent a good life grazing on green pastures. But as I joined the dots, a new understanding dawned. Intensive animal agriculture is shocking as far as animal welfare is concerned, but stopping it and switching to grass-fed beef and free-range chicken is not the solution either. Meat and dairy alone account for 14 per cent of global greenhouse gas emissions,[1] more than those from burning petrol and diesel for road transport. Of course, it is more nuanced than just carbon footprint. Meat and

1 From the work of the Food and Agricultural Organization of the United Nations (FAO) in assessing the environmental impact of livestock production

dairy are also not an efficient way to use our land for food. Their production utilizes three-quarters of all agricultural land, yet it takes almost 100 times as much land to produce a gram of protein from beef or lamb than it does from peas or tofu.[2] Not to mention that we seem to have a blind spot about the effects of livestock on our planet and on our own health. Animal agriculture, including poultry farming, has a frightening impact on deforestation, water pollution, antibiotic resistance and zoonotic diseases.[3] Beef farming is by far the main cause of deforestation in the tropics, since the farmers need land for pasture and for the soy required to feed the livestock. Here in the UK, our clichéd view of the countryside is of sheep and cows grazing a patchwork of green fields extending as far as the eye can see. In reality, however, this bucolic dream is more of a nightmare, illustrating the fact that our country is one of most nature-depleted in Europe and retains a mere 2.5 per cent of its original ancient woodland. Yes, it's great that there is a public outcry in the UK about our waterways being polluted with human sewage, but how many of us are aware that phosphate run-off from chicken farms is one of the causes of 'dead zones' in our rivers?[4] Given all this, how could I have had my head buried so deeply in the sand for so many years? I really had no excuse – the information had been there all along. As a scientist, I had been trained to look at the evidence and to ask questions, so I felt I should have known better. It was time to stop being part of the destruction.

---

2 'If the world adopted a plant-based diet, we would reduce global agricultural land use from 4 to 1 billion hectares' from ourworldindata.org

3 Zoonoses are infectious diseases that are transmitted between species, for instance from animals to humans.

4 Algal blooms reduce dissolved oxygen available for the growth of plants and fish through a process called eutrophication.

The solution is staring all of us in the face. We could stop the main cause of deforestation, significantly reduce greenhouse gas emissions and restore natural habitats[5] while feeding the world. Not to mention preventing millions of animals from being bred and slaughtered in horrendous conditions that we would never allow our pets to endure. What's not to like? So I started collecting vegan and vegetarian recipes and experimenting with new culinary ideas. I wanted to see how hard it would be to give up meat but still have a delicious diet. And the reality is that it is possible without sacrificing any of the pleasures of eating great food. Far from it, my diet has never been so varied and my taste buds never so stimulated. I really do not miss eating meat or drinking cow's milk. I have discovered so many delectable varieties of lentils, beans and peas, each with their own unique tastes and textures, and all providing great sources of protein. I have not turned my back on all the eating habits of my youth. I still use my childhood knowledge to eat seasonally, but I now embrace the amazing diversity of spices in Thai, Indian and other cuisines from around the world, especially to add zing to winter root vegetables.

Since I first arrived in the UK in the late 1980s there has been a complete transformation in what people eat and how they talk about food. Some may even call it a revolution. This should give us all great hope that a second food revolution is not far off. If a born-and-bred French carnivore can learn to love cooking nut roasts instead of roast lamb, then I'm pretty sure that anyone can.

---

5 I understand that some grazing can be beneficial for the soil and biodiversity.

WHY I NO LONGER FLY

Pete Knapp, PhD

Pete wanted to be a medical doctor but, with no offers from medical schools, he took a degree in maths with astronomy. After a four-year stint as a drummer, he trained to be a teacher in the UK and his first job was in China. His work then took him to Switzerland, which led him to take a master's in antimatter physics. A PhD in air quality followed, but the university's stance on activism made this a hard journey. He writes this just after completion of the PhD while he starts a career in journalism. His first major project is to make a documentary about European communities who are fighting against two mounting threats – fires and fascism. He reached six countries over three months by train, bus and boat.

Growing up in suburban England was a little dull, and global travel was a hugely exciting prospect. Inspirations from Michael Palin, Tintin and Asterix, as well as travel shows and relentless airline advertising planted many seeds into my impressionable mind: I wanted to see the world. Along with the popular sentiment 'you can do anything if you put your mind to it' that many of my generation and privilege were told growing up, I aspired to seek adventure. The ultimate freedom was to travel to the furthest corners of the world.

Once I had trained to be a teacher, the opportunity to travel more extensively came when I accepted a job in an international school in Beijing. I wrote a book about a trip I took to North Korea, motorcycled across Japan, and horse trekked through the glades and mountains of China.

I posted my adventures on Facebook, where I received social acceptance and praise, and was even told that I inspired others. It felt like I was riding on a limitless and unaccountable wave of success where the only way was up. I was never challenged about the environmental costs of this lavish lifestyle. I flew to four destinations every year and back to the UK every Christmas, producing 17–34 tonnes of greenhouse gases (written as CO_2 equivalent, or CO_2e). This is compared to the current 8.5 tonne annual average produced per UK citizen, and the 1.4 tonnes yearly limit calculated by the Hot or Cool Institute that we each need to meet by 2040. This think tank says that our CO_2e then needs to shrink to 0.7 tonnes per person in 2050.

But travel to the remote primary rainforest of Borneo in late 2018 showed me the reality and scale of human impact on the environment. From the window of the propeller plane, the entire stretch of rolling hills that could be seen was a palm oil plantation. What was once rich and lush rainforest, home to diverse and unique symbioses, was now a graveyard. The Bornean rhino was declared functionally extinct in 2015 due to loss of habitat. My conscience was pricked, and my bubble burst. The invincible wonders of the natural world that David Attenborough had presented had been ground into palm oil used in soaps, toothpastes, chocolate, bread and biscuits. This was a pivotal moment when my life took a sudden change.

Upon returning to Beijing from Borneo, the plane descended through a thick layer of coal-smoke like a submarine sinking into the abyss. I started to find the Beijing air pollution intolerable, all generated from economic growth, industrialization, cooking and planes. I started to see the health burden carried by people living near the airports or under the flight paths, who suffer the combined effects of air and noise pollution. Many students at my school suffered from asthma and could only participate in sport inside the school's dome, which was pumped with filtered air. One child had to wear a kind of hazmat suit because the air pollution had made him allergic to sunlight. Yet, I was one of the very few people wearing a mask to social events, cinemas and restaurants because of the air pollution.

I felt a bit weird, that there must be something wrong with me. Why was I the only person so worried about air pollution? I dropped international teaching and headed back to the UK to do a PhD in air quality. In academia, I felt I would be immersed in a culture that creates change and pushes the limits of what is possible. But, to my surprise, the institutions suppressed radical challenges and facilitated the death spiral of 'business as usual'. Compounded by the pandemic, I felt alone here too.

Shortly after starting my PhD, my grandmother died. Her funeral was the lowest moment I can recall. I didn't realize it at the time, but processing my grief over my grandmother was so unbearable because I was also grieving for my future, my identity and life on Earth. My awareness of the scale and urgency of the climate and nature emergencies exploded into clarity only once Extinction Rebellion covered the front pages of the news media in 2019.

I joined an XR march, where I discovered a very different atmosphere to what I had expected from the media coverage. People were friendly, willing to talk, and didn't fit the activist stereotype reported in the mainstream media. I joined Scientists for XR, and finally found the community of people I was looking for. I interviewed many of them for a podcast series called *Tipping Points,* where I asked them how they managed to withstand alienation from family, friends and colleagues. Their responses helped me to cope and also to develop my understanding of how air pollution, rising carbon emissions, the decline of nature and excessive consumption are all connected. One interview with an ex-pilot brought into sharp focus the inequality, injustice and scale of pollution caused by air travel. I couldn't hear this without feeling an intense responsibility and guilt for my past globetrotting, and seeing the strong links between the destructive fossil fuel economy and flying.

The UK ranks the third highest for total air travel emissions in the world, with the USA and China ranking first and second. Yet, the UK has a smaller population than twenty other countries. With the average passenger plane burning through four litres of fuel every second, living the '1.4-tonne per year lifestyle'[1] would mean no flying at all. It would also mean adopting a local and vegan diet, using public transport and repairing things.

I started to see the adverts for holidays abroad, cheap flights, travel books and travel programmes as 'selling a dream' at the cost of so many who would never fly and whose futures will be ruined. With my PhD research group's focus on the environmental impact of aviation, I felt that the level of greenwashing in the industry was obscene. All savings in fuel efficiency are wiped out by flight numbers increasing at around 4 per cent every year. Building and

1 The UK average is currently around 8.5 tonnes of CO_2e per year. According to Mike Berners-Lee, author of *How Bad Are Bananas,* this needs to be reduced to 5 tonnes as soon as possible, then down to 1.4 tonnes by 2040, and to 0.7 tonnes by 2050. The biggest reductions will be in cutting out flying, eating plant-based diets, changing to a green energy provider, improving home insulation, and ethical banking and pensions.

scaling up the production of electric and hydrogen planes before our global CO_2 emissions quota is spent is impossible, and fresh water used to make hydrogen is both rapidly diminishing and needed to grow food. If all planes switch to 'sustainable' aviation biofuels, an area the size of India would be needed to grow the crops, and cause even higher carbon emissions than fossil fuels due to deforestation and land-use change. Contrails, which reflect the Earth's heat back at night, are not included in net zero targets despite roughly doubling the warming from the CO_2 produced by burning the fuel. And finally, most carbon offset schemes are fraudulent, fail to reduce emissions, and remain colonial – the majority of projects happen in the Global South, resulting in local conflicts and land grabbing. Only 2 per cent of offset schemes have a high likelihood of success.

I feel now that flying is a pastime for the privileged at the expense of the poor. More than 80 per cent of the world's population has never taken a flight, although Boeing markets this as a reason to continue the growth of aviation. Just 1 per cent of people cause half of global aviation emissions. Talking to people about their next flying holidays became unbearable and I had to stop going to family gatherings where these conversational landmines were so abundant. At its most extreme, I see flight tickets as death sentences for anonymous members of our global society.

I pledged to go flight-free for at least a year in 2019, and I haven't flown since. Soon after, my sister moved to Canada with her young daughter. When she also made a pledge not to fly again, physical access to my family became impossible and the Internet signal was then so bad that it was distressing trying to speak with them. A return trip to see them would produce 2.6 tonnes of CO_2e, or 5.2 tonnes including night-time contrails, making living the '1.4-tonne lifestyle' impossible. It would add to the risk of wildfires that threaten their lives directly from the flames and indirectly from the smoke (the area of Canada burned in 2023 was more than twice that of any year since 1983). It would also add to the risks experienced by so many others around the world: it would add to the air and noise pollution around airports that is already killing people. I couldn't say that my need to visit my sister was more important than the life of someone else. It didn't feel that to stop flying was really a choice; I felt it was my duty as a privileged global citizen.

I found doing a PhD in the climate emergency incredibly difficult, especially when working closely with those researching global heating from aviation. I felt there was little recognition of our social responsibility as scientists. None of them appeared to see the contradiction in flying, even to conferences to report how bad flying is. So much research in the Global South is conducted by privileged people from Western universities who fly there, rather than providing local resources or training local people. Senior researchers fly the most and are often celebrated by their institutions for their international presence. What signal does that send? I also didn't hear anyone examining how their public funding makes them subject to net zero targets, or that a major share (up to 60 per cent) of universities' total greenhouse gas emissions is from flying. I didn't manage to convince anyone in my research community to join me in speaking to the London Gatwick planning inspectorate about the harm we knew would result from the airport's proposed expansion. I found no space to challenge the hypocrisy in academic flying: attempts I made to do so were met with hostility and in the latter stages of my studies had me banned from the university campus. While this felt very alienating, I was not alone. My friend Gianluca Grimalda, whom I know through scientist–activist networks, was fired by his German university for refusing to fly back from Papua New Guinea, despite his work being about how climate change is affecting indigenous groups there.

This is where I find myself today. I feel frustrated by the prevalent hypocrisy and neocolonial attitudes in academia, how it champions logic over ethos, and how poorly it speaks with and listens to the public. Now that I have my PhD, I have left academia to start a career in independent journalism where I feel I can be most effective to help to accelerate societal change. People are worried about two things that are on the rise in Europe – wildfires and support for the far-right – and my documentary explores how solutions to both may lie in building back lost communities, and restoring trust and connections between people. Other than saving 90–95 per cent of the carbon emissions of flying, my trip by train across six countries allowed me to feel more connected to the changing countryside, and to feel like I was really travelling. I even won a competition for making the documentary using public transport.

Global travel by plane had its benefits, of course. But the discovery of the world outside my head was just one half of the journey. The journey 'inside' my head, which was facilitated through activism, was long overdue and has allowed me to grow into a much more compassionate, aware and responsible adult. Who knows, maybe I'll find a way to see my Canadian family by boat, but until then I will make the most of the easy and instant messaging and video calls[2] that we didn't have just a few years ago.

2 A video call produces between 0.00015 and 0.001 tonnes of CO_2e per hour, meaning that one return flight from London to Calgary produces the equivalent of up to 37,000 hour-long video calls – one per day for 100 years. Video emissions could even be reduced by 96 per cent with the video turned off if you're having connection issues or a bad hair day.

A PERSONAL PERSPECTIVE ON POWER

Isabella Stephens, FHEA

Isabella is a PhD student studying sustainability in batteries. This means asking how the materials chosen for making a battery affect its sustainability. She has a master's in chemistry from the University of Oxford, and is hoping to use batteries to create a more equal world. She has struggled to disentangle 'science' from the social and exploitative nature of our world. She tries to think more holistically, where lab research to make better batteries is necessarily paired with community-building to create the better world in which those batteries will be so needed.

My route to activism was through walking. Walking around places built for cars and feeling threatened by the hostile setting. I knew I belonged on the pavement. I knew I could and should walk to work: it was less than 2 kilometres. Yet why did I feel so at risk? Why did the pavement disappear into the wall, when walking is one of the most accessible, healthiest and cost-effective ways of getting around a city?

As humans, we have several things that unite all of us. One of them is movement. Travelling from one place to another, however short or long, is something every person experiences. You might walk to the shops, or you might get a train from London to Edinburgh, or maybe go further afield. Our bodies have evolved to be excellent at long-distance endurance travel on foot. So for me, it is one of the most natural areas in which to start taking action for the planet and against inequality.

I grew up in a fairly rural part of the south-east of England. The local bus service was pretty terrible, and I was dependent on my parents driving me places, but close to home I happily roamed the fields on footpaths completely detached from roads, walking in close harmony with nature. Moving to Oxford for university, I suddenly was able to get myself everywhere on foot, and felt liberated. I actually moved a lot less in terms of distance, but moved my body a lot more, since I wasn't ever getting into a car. My world radius shrunk but my possibilities grew. Oxford was dominated by pedestrians and cyclists. Sometimes the cycling was a bit sketchy on the bigger roads, but I never fundamentally questioned whether I should be moving through the city in this way. I'd be crazy not to cycle around; it was easily the fastest way for me as an able-bodied person to get anywhere.

Pretty early on into my undergraduate chemistry degree it became clear to me that I wanted to work on decarbonization. I had always been concerned about climate change, and was really interested in energy. For someone studying chemistry, this meant nuclear, which seemed slow and contentious; solar, which had already had its big technological breakthrough; or energy storage. We could already generate abundant clean energy, but in order to have an energy grid that is truly free of fossil fuels, the problem of energy storage still remained a challenge, so I chose that area. I was particularly captured by the idea of lithium-ion batteries as the future for cars and for home stationary storage (for example where houses have their own battery packs to store excess solar energy or charge during cheap tariff hours). Maybe this was because I held a battery in my hands every day, in my phone, and I could tangibly see how batteries could be a real part of the future.

At the end of my degree, I couldn't think of anything worse than doing a PhD. I am someone who really needs practical applications and an end goal to motivate me, and the thought of exploring fundamental science with no real application did not appeal. However, I wanted to learn as much as I could about making batteries better, and with a lack of jobs in what is still a young European industry, as well as me not really knowing exactly what I wanted to work on, I was recommended a PhD that would look at sustainable battery materials. I applied on the off-chance.

My PhD took me to Birmingham in the English West Midlands. The UK's second biggest city (a title hotly disputed with Manchester depending on exact parameters) is full of life, with lots to do and much more reasonable living costs than London. Naively, I assumed I'd be able to cycle around as I had in Oxford. The places I wanted to go to were distances I wouldn't bat an eyelid at, nothing more than twenty to twenty-five minutes cycling away. The first few times I cycled on the roads I felt like a piece of dirt. I wasn't used to drivers having such utter contempt for my life and right to be on the road, and often would arrive at the university shaking after a driver had driven really close past me at speed. There was no off-road way to get to work, and walking took more than twice as long as cycling. I didn't want to give up half an hour of my day and my freedom to travel because small-minded strangers wanted to speed in their massive cars on tiny Victorian streets and became angry when you made them go a bit slower. They used the roads around where I lived – streets where (in general) those drivers didn't live – as places to shave minutes off their journeys. Add into this situation that, in the UK and globally, communities that have the lowest rates of car ownership are exposed to the worst traffic-related air. The vast inequality that the car-dominated transport system promoted was quite literally choking.

I changed the way I cycled to be more assertive and take up more space, so that drivers couldn't squeeze past on narrow roads. I accepted the grim reality that the only things that would keep me alive were my own vigilance, assertiveness and good luck. I started to question the transport system that I'd grown up with. Why did one person driving to work deserve to take up so much space and resources? Why did cities not incentivize much more efficient ways to move around such as public transport, safe walking and cycling? Humans are social creatures, and yet cars divide us into our separate boxes and make us much more isolated. Why did we seem to prioritize cars so much?

While feeling as if I had a twice-daily assault on my safety and liberty, I was at the beginning of my PhD studying cathode material for lithium-ion batteries. So much of the justification for these batteries was for electric vehicles, but I was rapidly falling out of love with cars and waking up to the reality of a world dominated by their impact on our built environment and our bodies and minds. My first PhD piece of work centred around nickel materials for batteries. I read more and more evidence that nickel production was linked to deforestation of rainforests in Indonesia. Cobalt production was linked to child labour in the Democratic Republic of Congo and lithium production to violations of indigenous water rights in South America. Was all this really any better than fossil fuels?

I wanted to make batteries that made the world better and more equal. But most of these batteries were going to go into cars, potentially large cars, and were fuelling the systemic transport problem. I struggled with the idea that this might not be the right way to build a better world. Larger cars like SUVs cause more damage to infrastructure than small cars, since road damage is directly related to car weight, and no wonder cars don't fit in parking spaces anymore when they have been growing a centimetre wider every two years for the past two decades. SUVs are also much easier to flip over due to their higher centre of gravity, making them more dangerous for both the drivers and people around them.

The answer is not as simple as electrifying everything. In order to electrify the world we live in, we would likely destroy what we are trying to protect. The consumption-focused world must also change, meaning that the way we move around it needs to use resources more efficiently. Cars sit still for most of their lives, and yet so many people want to own one themselves. As these dots started to join in my head, I began to give talks about systemic problems in transport and climate justice, and how I saw it all intertwining with my PhD in material science. This was a painful realization. It was not enough just to do good science and make batteries better. What if the answer was instead to radically change the way we live? To move in a way that connects us to the places we travel through and to our own bodies? My life and work would have to encompass both of these elements. Every battery would need to be used in a way that made it worth the sacrifice of digging those minerals out of the ground, while also improving mining practices. So I joined communities that felt the same way, and started to put my energy into building this better world with them.

During outreach activities, when I have my lab coat on, people often ask me about lithium since there's a lot of articles about this problem. 'Do we have enough lithium?' is one question that always makes me smile. The answer I tell is that, 'Yes, we do, but not if everyone wants an electric SUV.' That's the nature of this problem: we won't win with an individualist approach. We need batteries to power vehicles and give our electricity grids the resilience to rely on just the wind and Sun. However, the obsession with larger and larger cars threatens to gobble up some of the few remaining intact ecosystems in this world in its insatiable desire for minerals like nickel. Some people will always need cars, but we need to reconfigure our transport, particularly in urban areas, to move people efficiently and make each mineral ion mined out of the Earth do as much as it can.

Imagine a world where everyone can get from one place to another quickly, easily and without causing harm to anyone, human or nature. That's what we need to build, rather than electrifying the car-centric world.

Dream big.

A SEARCH FOR SUSTAINABILITY, FROM THE FORESTS OF MADAGASCAR TO THE HILLS OF WALES

Ryan Walker, PhD

Ryan is a conservation biologist. He undertook a PhD through the Open University researching effective methods for surveying rare, difficult-to-detect, critically endangered species. After ten years of fieldwork in Madagascar and Papua New Guinea, he now works occasionally as a self-employed commercial ecologist in the UK. He lives on a smallholding in Wales with his wife, and spends most of his time managing the land for biodiversity conservation, growing food, making charcoal and biochar in the woods, and building useful stuff out of other people's junk. He is also involved in environmental activism.

For as long as I can remember, the natural world has fascinated me. All I had ever wanted to do was work on the conservation of threatened species in their natural habitats. Ten years ago, I managed to achieve this. I was lucky enough to be a freelance field biologist for a number of conservation NGOs in Madagascar and Papua New Guinea, leading and mentoring small teams of local biologists. I was in two of the most biologically and ethnically unique and diverse places in the world, working with critically endangered species and with local people who aspired to be the conservation practitioners of the future. I really had managed to get to a place in my career and life that I had always aspired to, ever since I was an eight-year-old collecting and identifying spiders in the back garden. Or so I thought at the time.

For the best part of a decade I lived a very transient life, flying between Madagascar, Papua New Guinea and the UK on an annual basis, and spending most of my time in the field in isolated places. The initial enthusiasm during the first few years made me feel like I was making a positive difference. Two local PhD students and I had mapped the extent of the remaining range of two critically endangered tortoises in southern Madagascar; two species that were almost unknown in terms of their ecology and biology, but were literally disappearing before our eyes due to poaching and habitat loss. In Papua New Guinea, I was proud that in the space of a few field seasons I had helped take a small team of local university graduates from enthusiastic amateur natural historians to effective, professional field biologists.

Then disaster hit southern Madagascar. Already one of the most marginalized and poor communities in the world, that has historically staggered from one crop failure to another, the region was subjected to what the United Nations described as the first climate change-induced famine. Predictably, wildlife poaching and forest loss increased as people scrambled for whatever resources they could find to avoid starvation. Meanwhile, in Papua New Guinea an unintended consequence of my work with the young local biologists was becoming evident. In the space of a couple of years, the mining industry had swept up most of the team that I had been training and working with. Local legislation dictates that a cursory attempt at an environmental impact assessment needs to be undertaken whenever large, international mining companies prospect for and exploit the extensive gold, copper and other metal reserves that Papua New Guinea holds in abundance. To do this, mining companies need teams of environmental scientists including field biologists, and what better recruiting ground than the lowly paid conservation NGO sector, with huge salaries effectively luring people away? I was very fond of the team that I worked with and trained; they were all good people. But cultural responsibility in that country means that everyone with a professional salary has a duty to pay school fees for younger siblings, cousins and other members of their extended family. Then there's medical costs for older, ailing relatives; food for other family members falling on hard times ... the list goes on. If you are relied upon as the local social security system, moral judgement is a luxury that you often can't afford, especially if you are paying for the malaria treatment for ten kids in your family network.

I was feeling a little dejected by this point in my career, with the ever-growing realization that I may have indirectly had more of a positive impact on the shareholder profits of a few Chinese mining companies than on the protection of the natural resources of Papua New Guinea. Added to this was my guilt over the amount of carbon-belching air miles I had racked up, when southern Madagascar was a living case study in the devastating effects of climate change. I decided to go home. I got married and spent some time thinking of what to do with myself for the rest of my life.

As an undergraduate environmental science student in the late 1990s, I was well briefed on the challenges of recent global environmental issues and how society came together to rectify these problems. Whether it be improvements in air quality to tackle the problems of acid rain on the northern boreal forests of Europe and Asia, or the global banning of chlorofluorocarbons to protect against ozone depletion, society seemed to be able to sort it out. My fellow students and I also learned of the devastating potential impacts of a warming climate as industrialized countries burned ever more fossil fuels. However, even in the 1990s, society appeared to me to be heading in the wrong direction on this, by allowing more carbon to be emitted into the atmosphere year on year, despite the rapid development of clean sources of energy. Repeatedly over the subsequent years, I thought that surely this is a problem that can be fixed?

During my contemplative period, whilst questioning the effectiveness of my career in conservation, a new group of environmentally conscious, motivated, ordinary people appeared to be mobilizing and asking that very same question. How have we got to this? How could a problem that we have known about for more than forty years be allowed to get progressively worse? How have we reached the stage where it is now killing and displacing millions of people and devastating biodiversity, just because fossil fuel companies, other multinational capitalist interests and a few phenomenally wealthy individuals have become more powerful than the governments of the world? Extinction Rebellion really caught my attention. I was particularly drawn to its promise of mass civil disobedience, based on effective models such as the American civil rights movement, to force the UK government to implement policies that will protect us from the effects of climate change, and to the idea of people's assemblies to give environmental decision-making back to the people and away from corrupt officials.

So, there it was. Maybe sitting in the road in London with thousands of others could do more to protect the biodiversity that I cherish than working as a conservation biologist in volatile places in the world. Even better, I soon discovered, was sitting in a road with a white lab coat on as a member of Scientists for Extinction Rebellion. I could tell anyone who wanted to listen about my first-hand experiences of witnessing the horrors of climate change-induced food insecurity, the devastating effect it has on humans and wildlife, and how we all have a responsibility to do something about it. The starving people of southern Madagascar have no voice – they deserve to be spoken for by somebody, somewhere, I thought!

During my time working in remote regions of the Global South, I was always fascinated and impressed by the skill and ingenuity of people to get by on very little. There were elaborate dwellings made from scrap and local bush materials, basic hydroelectric and solar systems, rainwater collection and irrigation techniques, and small-scale agriculture that takes advantage of changing conditions. In Papua New Guinea, there is a strong tradition of sustainability based mostly on necessity, as a result of a firm belief in local resource ownership. So for example, if some people overfish the reef in front of their village, the community a mile down the beach will not donate any of their fish from their reef. A 'your reef, your fish – fish it properly because we are not going to go hungry through your mistakes or short-term greed' sort of attitude. This makes a lot of sense, even when you extrapolate it to a global scale. There would be enough for everyone if we all respected planetary boundaries – it's just that some take more than they should.

I was buoyed with the enthusiasm that Extinction Rebellion was now out there mobilizing thousands of people like me for positive change, but also frustrated by the injustice of what was happening to the people of southern Madagascar and impressed by what I had learned from the progressive attitude of one of the oldest cultures in the world. My wife Beth and I decided that we would, from the privileged position of living in one of the most economically developed counties in the world, try to live as lightly as possible.

We decided to move back to south Wales, where Beth grew up. I'd been watching the cost of small patches of land in southern England for years and realized it was out of the reach of most regular people. However, Welsh land prices seemed more achievable. Fully aware that just buying a patch of land, sticking a caravan on it, constructing a vegetable patch and propping up a couple of solar panels is far, far more difficult in the UK than in most places in the world, we still pressed on with our plan. The UK has some of the most stringent planning regulations in the world and this could be considered a good thing, since we live in a densely populated country. However, when you delve deeper, you soon realize that there is a lot of slack in the system if you own a huge, industrialized farm, or have enough money to buy a modest rural property, knock it down and build a mansion. Smallholders and small-scale land workers have been forced off the land incrementally for hundreds of years in favour of huge estates and environmentally damaging agribusiness. Maybe the model of direct action also needed applying elsewhere?

We eventually found a twenty-acre patch of land, comprising a mix of overgrazed meadow, self-seeded deciduous woodland, ancient woodland and the dilapidated remains of a market garden. However, most importantly, the previous owners had lived on the site for ten years in a caravan, un-harassed by the local council's planning enforcement department. We pooled all of our resources and bought it. Our aims: to build a small, discrete, comfortable home, using as few resources as possible to build and operate; produce as much of our food and other day-to-day resources as possible; set up a small, sustainable land-based business; and manage the land positively for biodiversity conservation.

None of what we are doing is in any way unusual or radical, given that most people globally live a subsistence lifestyle, reliant on local resources and a robust food system of tens of thousands of cultivated species, mostly of local origin. This compares to the few tens of species used in modern, globalized, industrialized agriculture that we mostly rely on in the UK, with all of the environmental and social impacts this entails, as well as the vulnerabilities to disease and rapidly shifting climatic conditions. However, as a result of the Town and Country Planning Act, every day is an act of defiance if you try to live quietly and sustainably on the land in the UK. You are challenging a capitalist model telling you that you need to spend a lifetime in debt to have a home, when with a bit of ingenuity and patience, a decent, comfortable, functional home can be built for less than £20,000. That same capitalist model wants you to believe that you need strawberries flown from Argentina in December, when growing good, organic, winter root vegetables isn't difficult but just time-consuming ... and that time has been taken away from us if we're exhausting ourselves working to pay off that excessive mortgage.

These days my priorities are focused on forging links with other local smallholders and off-gridders as we share knowledge, skills, labour and equipment. Local networks are important as we move into more uncertain times. I feel it is important to try to normalize a pragmatic, simple and sustainable way of life that we have become so far removed from. Similar to the Extinction Rebellion model, if enough of us do it, it becomes harder and harder for anyone to stop us.

SAYING GOODBYE TO THE UNIVERSE: THE EPILOGUE TO MY DOCTORAL THESIS

Lucy Hogarth, PhD

Lucy has a master of physics degree from Durham University and a PhD in astrophysics from University College London. Lucy is neurodiverse, having both autism and ADHD. Despite those barriers, she joined Extinction Rebellion after she started her PhD in 2018 and was struck that she'd come that far in academia with no one mentioning how critical the climate crisis was. Doing a PhD with a focus so far away from our world became increasingly difficult. To offset some of that dissonance, Lucy became an active member of Scientists for Extinction Rebellion. Lucy has also designed a large collection of art for the XR movement. Seeing her work used all over the world is one of the most gratifying things in her life. At present, she has stepped away from astrophysics to explore how she can use her artwork to reach more people.

The following text comprises the final chapter of my doctoral thesis entitled *Going with the Radial Flow: How Molecular Gas Flows Affect the Large-Scale Evolution of Galaxies.*

'If people insist on living as if there's no tomorrow,
there really won't be one.'
– Kurt Vonnegut

Beginning this thesis with the stunning Lena River Delta in the far east of Russia was not an arbitrary choice of metaphor for our Universe. The structure of our Universe is echoed throughout the complex structures and systems on our planet; each perfectly balanced within its ecological niche and tuned to its environment.

The Lena River Delta is a fantastic analogue for the cosmic web, not only visually, but also due to the rich wildlife and diversity it supports through its dynamic processes, not unlike the wealth of galaxies fuelled by the cosmic web. However, unlike our cosmos, the Lena River Delta is rapidly deteriorating as human activity continues to alter the environmental constants that support life on our planet. Vast cosmic simulations have demonstrated the degree to which parameters and constants have to be fine-tuned to produce the cosmic ecosystem we observe in our Universe. As scientists, we know that tiny changes in the parameter spaces of these complex simulations can have dramatic consequences as they evolve. Our planet is in a balance that is just as delicate and equally as tenuous.

In a thesis exploring the properties of nearby galaxies, the idea of looking down towards our own planet seems entirely at odds. However, feeling 'entirely at odds' is precisely how studying for and writing this work has felt for me. In the five years I have been a member of Scientists for Extinction Rebellion, I have stood in front of thousands of people and told them how 'I was looking to the stars, while the Earth was burning under my feet.' Every time I said that phrase, it has pained me a little more to go back to looking up at the Universe.

I have dreamed for so much of my life of the privilege of studying the cosmos and I have had to fight for it. So many times, training to be where I am now, writing these words, has brought me closer to the edge than any person should ever be. Studying the Universe is a privilege, though, and with all I know about the consequences of the climate crisis, many of which are already manifesting across our planet, I find it increasingly hard to see it as a privilege that humanity can afford in this moment. I believe, as scientists, our actions carry a special weight in the eyes of the public, regardless of our field. By continuing as we are, not acknowledging our responsibility or need to change, we fuel the stagnancy that inhibits true action against the climate crisis.

Scientists are motivated by a relentless and scrupulous passion to pursue truth in their fields of study. It was that passion that drove me here, but I am unable to compartmentalize that passion to just the work contained in this thesis. I truly believe that without us acting holistically as scientists, with a unified action on a topic that affects all of us, science will be irreparably damaged and so will our planet.

I never wanted to end this thesis this way; I am where I always dreamed to be. The dissonance I feel though, when I'm looking up when I know I should be looking down, is too much for me to bear, and for now I cannot continue my work as a scientist. Increasingly, whenever I look upwards, towards those unique and gorgeous galaxies studied in my thesis, I come back to the same question:

'Am I looking up, or am I looking away?'

WHY I KILLED MY CAREER

Alison Green, PhD

Alison is a cognitive psychologist and expert on human learning and thinking. Her academic posts included Dean at the Open University and Pro Vice-Chancellor at Arden University. At the end of 2018, she traded academia for activism and resigned from her post, which was reported in both the Guardian *and the* Financial Times. *Alison is currently Executive Director of the Scientists Warning Foundation. She has written about the role of academia in perpetuating the climate and ecological crisis and about climate change engagement of scientists. Alison is an active campaigner, working with numerous organizations and groups, such as the rewilding campaign group Wild Card and most recently WePlanet's Reboot Food campaign*

I remember the first day so clearly. A cold and clear November morning in Cambridge, and I was about to board a train to London to do something so out of my comfort zone. I couldn't quite believe that I was doing it. Most times I journeyed to London, I'd be in a suit and heading for a board meeting at our academic headquarters or at one of our campuses, in my role as Pro Vice-Chancellor of a new and busy university. But today I was heading for Waterloo Bridge, unsure about who I'd find there and what was going to happen. As I stepped off the train in London, I tried to prepare myself for what I was going to do – when the call came, dash out onto a busy bridge in London and sit down in the road and bring the traffic to a standstill. I wanted to be there because I am a scientist, and because scientists' warnings on the environment are not being heeded. My mouth was dry and my heart lurched as I made my way to the bridge, scanning for anyone else who might be about to do the same as me. What do environmental activists look like? I had no idea. Everyone seemed so normal. I couldn't quite believe that in a matter of minutes, I was going to jump over the barriers, run into the road and sit there. I approached one person on the bridge and clumsily asked him if he was there for the same reason as me. 'Let's just forget that this conversation happened,' he replied with a caustic grin, walking away. I was mortified.

Yet from that day on, nothing was ever the same for me. I sat with scores of people on that bridge, most of us strangers yet each there for the same reason – profound concern about the unfolding planetary crisis. I had to abandon my usual reserve and was glad to find that talking with others was easy. I chatted with an elderly couple who were there because they wanted their grandchildren to have a liveable future. I spoke with someone who had travelled thousands of miles to be there, believing that activism was now our best, and perhaps only, chance to try to drive action that would ultimately bend the greenhouse gas emissions curve down. I felt calm and relieved that I'd been able to take this action. Maybe doing the right thing isn't so hard after all?

I grew up mainly in Scotland, on the Moray Firth, and then went on to study psychology at Aberdeen University. I became an expert on how we learn and develop expertise, and found time for more applied work, too, even spending a couple of weeks on an offshore installation in the North Sea, surveying the attitudes of offshore workers to the future of North Sea oil and gas – and that was revealing. A number of the men completing the survey came to me and spoke of their concerns about chemicals and substances being dumped overboard if they were spoiled. Yet when I rather naively commented on this to the platform managers, they denied that it ever happened. It was for me an early lesson in the ways in which the process of becoming a manager can at times skew thinking to blot out or deny practices that could be frowned upon.

One of my interests has long been in how research informs practice, and the ideal that many of us hold dear: that the experimental and theoretical work that we do as scientists is then taken up and used for the benefit of all. It is an aspiration that is at the heart of the 'science–society' contract. Many academics are motivated by this same desire to do research that yields tangible benefits.

However, for months leading up to the bridge action, I had been tormented by the stark realization that in spite of decades of warnings from scientists about a looming planetary emergency, we were still on track for what some scientists have called a 'ghastly future'. Driving to work each day, just another casual commuter hurtling along relentless ribbons of sticky tarmac, our looming predicament felt like a dead weight in my dark and depressed mind. I watched the Sun and Moon rising and setting, overwhelmed with the guilt of my species. My research area was full of examples of how experts are valued for their knowledge and problem-solving prowess. So how could it be that climate scientists and conservationists, who had been diligently documenting for decades a worsening of a number of 'planetary health' indices, were not being listened to? What was going wrong?

Worse, far from being a part of the solution, I now saw very clearly that academia was part of the problem. I had long believed that education was a public good, and I began my academic career simply wanting to help others to learn. Most researchers engage in research with the greatest integrity, and it is undeniable that we are largely better off for the outputs of that research. But as I came to understand the drivers of the planetary crisis – the relentless push for profit and 'progress', and over-consumption by a minority – I realized that academia as a whole was in fact inadvertently preserving the status quo rather than being at the vanguard of change. Neoliberalism has captured university education in the UK, leading to the commodification of education. In short, academic institutions have become businesses, their focus very much on the bottom (financial) line. This drift has been gradual and it has been pernicious. Less popular courses in areas such as arts and humanities may be axed or come under intense threat of being axed. This is not what education should be.

I had learned about some startling new research exploring the extent to which fossil fuel interests have permeated higher education. These influences range from funding research that supports the fossil fuel industry, to influencing the content of academic courses, and even to being present at recruitment fairs. In a context where the International Energy Authority has stated that there must be no development of new oil and gas if we are to meet the Paris goal of limiting warming to 1.5°C, then it is nothing short of outrageous that fossil fuel interests are using academic institutions to encourage young people to go into careers in oil and gas, careers that have no viable long-term future and exacerbate the planetary crisis.

I thought long and hard about what to do and talked openly with friends and colleagues about my dilemma. How could I continue to work in a sector that had somehow become a part of the problem? I thought about the many hundreds of students I have taught and engaged with over the years, people embarking on university courses in good faith, with hopes to become good citizens and to make a positive contribution to society. I thought about the courses they were being offered, courses that appear to assume that the future will look much like the present, a highly doubtful premise. And I thought about my three daughters, all of them either at or going to university, and my responsibilities to them. Could I use my voice more effectively within academia, or outside of it? Those I confided in were overwhelmingly supportive and understanding. Days after the bridge action, I resigned from my job as Pro Vice-Chancellor and embarked on a journey I could never have foreseen.

Fast forward a few years from my rebirth as an academic activist, and I find myself once again on a train to London. This time, I'm going to support Scientists for XR, who have a pop-up stall at an event outside the Science Museum. We're there to talk with the public about the Science Museum's choice of sponsor for its new exhibition on the climate, which is none other than Adani, a large fossil fuel conglomerate based in India. I walk towards the stall to meet the team, aware that there is a police presence in the vicinity. I seldom venture into London now, though I wanted to help out on this occasion. I have someone who is dependent on me, and so being arrested is not an option for me. My admiration and respect for the scientists fronting the campaign, wonderful people who routinely put their liberty on the line, is immense. I stand up to speak and feel nervous, because I can see a policeman in the crowd, quite close to where I stand. He is talking amiably with one of the scientist–activists in the team. I feel reassured by this and get on with what I want to say to the public, who are gathered around. It's a peaceful event and people seem interested in the points that are being made, such as calling out the Science Museum for accepting sponsorship from a fossil fuel organization for a new gallery on climate change. For a public increasingly concerned about climate breakdown, having scientists and activists work together to reveal greenwashing, disinformation, denial and all of the delaying tactics used by the polluters is important.

Several years down the line, I don't ever have to ask myself if I did the right thing in killing my career to join forces with scientist–activists and environmentalists – I know I did. I'm grateful for the privilege that made it possible for me to do that. I like the irony of finding myself better known now in the academic community as an activist than I was as a conventional academic, but I'm more interested in motivating others to act than I am in notoriety. Could I have been as effective had I remained an academic? I somehow don't think so, as I'd never have found the time to lead and support so many campaigns, campaigns that are showing signs of success. One key lesson I have learned though is that many, many people are concerned, but are unsure about what to do, or even how to do it. Stepping out of our comfort zones can be difficult. But the planetary emergency has already flung many into desperate situations, and we all face a future where precarity will be the norm. It should be no surprise that Michael Mann, one of the world's leading climate scientists, has described this narrow window of opportunity as 'our fragile moment' – the actions we take now really will determine our future.

EVERY JOB IS A CLIMATE JOB

Kara Laing, PhD

Kara completed her PhD in applied mathematics just as the millennium turned. She chose to transfer to engineering as part of a computer modelling team. Her main focus has so far remained on building models of vehicle crashes, interpreting them, and using that to help manufacturers make their vehicles safer. In 2019, Kara joined Scientists for Extinction Rebellion and her local XR group, as an act of stubborn optimism. She runs a non-XR group for climate-concerned engineers and is also a point of contact for XR Engineers. She participates in Scientists for Extinction Rebellion actions whenever she can and has provided backroom support for others.

In an interview for air quality researcher Pete Knapp's *Tipping Points* podcast series in 2021, I chose to be frank: I explained that I didn't talk about climate change at work because it was too scary. Working in analytical engineering, I am part of a team that advises companies on how to improve the mechanical properties of products they wish to sell. Effectively, we are trying to preview the future and mitigate any potential harms – but all under the illusion that 'business as usual' will continue, if only we make all cars electric. Shortly after the interview, I watched as friends in activism risked arrest and all that could mean. I started considering whether I could carry on apparently supporting the illusion of an impossible future, or whether I should change my job. Four years later, I'm in the same role. So what happened?

When I submitted my PhD, I decided that the best use of my skills was to work in road safety. I started work in a consultancy that supports car manufacturers in ensuring their vehicles are 'safe'. It was a good time to get involved: car safety had become the latest buzzword for marketing new cars, so suddenly manufacturers were moving their own goalposts. This started because the European New Car Assessment Programme (NCAP)[1] had begun to publish star ratings for cars and suddenly people wanted to buy vehicles that would keep themselves and their families safe, should the worst – a crash – happen. Manufacturers who worked to make the safest cars marketed the scheme. It's worth noting that this increase in public information has resulted in a step change in crash safety: as people have been enabled to make an informed choice, they've used their purchasing power. It's clearly been beneficial: crash deaths in the UK have halved. The most recent consumer criteria have successfully motivated car producers to make their vehicles safer for vulnerable road users, too. The NCAP approach is an elegant example of scientific methods being applied and leading to measurable societal benefits. Robust mathematical modelling means that we can investigate many possibilities and extract a range of possible outcomes for each safety test. Results from our models usually predict physical crash test results well enough that most manufacturers now do

1 www.euroncap.com

little physical testing, instead relying on the analysis. I hope, and even dare to believe, that I have made a difference.

So when I read the IPCC's special report on global warming of 1.5°C in 2018, it floored me. I had been complacently believing I was driving positive change, but it turned out that everything I was doing was inconsequential. I could see what the future was bringing to my child and what it was already doing to thousands of children like him. Figure 3.4 of that report hit me hard. It's a chart showing projected increases of global temperature and extreme precipitation. At my core, I'm a mathematician: for me, a graph can be strongly emotive, like music or a poem can be for someone else. In my work I had specialized in translating abstract numerical results from massive computer models into potential human impacts. But as an engineer, I was supposed to be helping to make the future better, and by the time of my interview with Pete, I felt I couldn't. It was hopeless; I was hopeless.

Community is essential for each of us, however we find it. In 2019, Extinction Rebellion put me back together, as my local XR group and Scientists for Extinction Rebellion gave me a way of channelling my engineer's need to make things better. Still, over the next two years I avoided talking about this in my workplace and felt unable to talk about my life as an activist – the thrust of that *Tipping Points* interview. There are a number of papers suggesting that an active role as a campaigner can actually bring credibility to climate, environmental or health scientists,[2]

2 Bhopal, A. (2023) '"Are you a researcher or an activist?": Navigating tensions in climate change and health research' in *The Journal of Climate Change and Health*; Gardner, C. J. *et al.* (2021) 'From publications to public actions: the role of universities in facilitating academic advocacy and activism in the climate and ecological emergency' in *Frontiers in Sustainability*; Thierry, A. *et al.* (2023) '"No research on a dead planet": preserving the socio-ecological conditions for academia' in *Frontiers in Education.*

but I wasn't confident that this would hold true in corporate engineering environments. That conversation with Pete in 2021 started me on a path of realizing that my employment and my ethics were at odds. Despite working in a job I loved with people I valued, despite having recently attained the role I'd been working towards, I started looking for other work. After all, how could I reconcile the work I loved with the change that I saw needed to happen?

In preparation for making the workplace move that I believed was necessary, I started to 'upskill' beyond my main role (all the free courses I could find and fit in!) and I mentioned 'sustainability' at the slightest excuse, namely ensuring the products we were working on would not deplete natural resources or disrupt the balance of our biosphere. It turned out that this was the right moment again: clients were starting to recognize the business case for making their products greener, and I was ahead of that curve. I became the person that the management of our small firm started to ask to look into sustainability options. I found it hard to talk about the activism, but easier to raise the issues behind it.

I found, or was introduced to, other individuals in the industries I connected with, who were on similar paths. Conversations were had, ostentatiously sometimes. 'You should talk to ...' happened more and more, making it even easier to start those conversations. Recently, I started a regular cross-industry meeting of engineers with two initial intentions: to create the safe space I myself needed, and to start to share ideas. These meetings are safe to put in my work calendar, so are visible in it. I can talk about the meetings at work, and I've invited my line manager to attend. At these meetings we sometimes talk about how we feel, sometimes talk about how to approach technical challenges, sometimes simply celebrate engineering ideas we've heard about or seen. For example, one engineer's company is expanding: it is a plastic reuse company, recycling at the grassroots level. Effectively people 'borrow' the plastic item when they need it. Another example: using rainwater to flush toilets is something another engineer talked about a while ago – I saw it in practice this week when I visited a large tech company's offices. Recently, a third engineer talked about an approach to solving the 'wicked problem' of how to transition both energy and waste flows, which is already helping me rethink all kinds of technical challenges. We share ideas on how to make product design more circular, so that the material used to make it can also be used for something else, or be reused or recycled. We talk about plastic a lot – incineration seems nonsensical when it can be reused in many cases, if the right choices are made early enough.[3] We talk about how changes in one sort of development can be carried over into others. A recent book on regenerative design in structures is even seeking to repair any environmental damage that has been caused in the past.[4] Despite being online, despite being informal, these get-togethers are the start of a community. We all know that we are not the only engineers seeking a better way of doing things. Progress is far too slow, but it has started to permeate into the mainstream.

It turns out that other people have been taking a similar approach in their own spaces. Like mycelium, we are creeping across the engineering biome. Opportunities to connect and share are increasing. Some kind of tipping point has recently been reached – for example, in the course of one week I received two separate invitations to attend another, older group, which I will definitely join. It feels like magic: we didn't know it, but we were

3 After all, 'heat from waste' is mostly greenwash of the most verdant hue when it's burning highly processed oil.

4 Broadbent, O. and Norman, J. (2024) 'The regenerative structural engineer' from the Institution of Structural Engineers.

actually never alone. As the number of engineers who are explicit about their concerns for our climate and biosphere grows, so do the conversations about how to make things better through our work. At present, the challenge of making my work part of the solution is hard. This is where my own community of similarly motivated individuals is essential and regenerative. Not only does it bring hope and inspiration, it also lets me hold myself to account. It helps me see what can be done and hear what my colleagues working specifically in sustainability say is needed next.

It isn't all done and dusted, but I've found a lot of people across the engineering industry who have been making the case for more sustainable product conception for a long time. They have generously made my personal transition much easier than their own:[5] how on earth did I not meet them before? Meanwhile, a number of people in sustainability careers who I now talk to are stepping out of their corporate roles, or coming to understand that they were employed only to tick a box. This gives me cause to pause and reflect. Comparing the progress we are making with that which we need to make, progress is inadequate: the UK's Committee for Climate Change does an excellent job of keeping track of opportunities and progress to meet our national targets,[6] but I do not see similar targets passed to producers outside the UK or importers of 'stuff' that is brought to the UK. I

5 For example, the Global Association for Transition Engineering, GATE, was founded formally in 2014.

6 Whether these are even appropriate is debated. See, for example, Hickel. J. (2020) 'Quantifying national responsibility for climate breakdown: an equality-based attribution approach for carbon dioxide emissions in excess of the planetary boundary' in *The Lancet: Planetary Health*; and Campbell, I. R. (2021) 'UK's share of the global carbon budget will be used up in 3.3 years' in *The BMJ*. As we haven't seen significant change in British habits or attitudes, we are now, in 2025, extremely close to exceeding our fair share of emissions.

do not yet hear of new requirements for all the products that are mentioned routinely at my workplace. Urgently, an ethos needs to be embedded through every part of the engineering profession, through anyone who plays any part in designing and producing a product, so that the harm of new production is mitigated or even reversed. What we need is for every job to be a climate job![7]

Slowly, change is happening in the organizations we work for. The push seems to come not just from consumer demand or the desire to not be called out as 'worst in class', but also because of the efficiency benefits of products that use less materials and energy. Some companies are changing themselves completely, some are improving a bit, and most are at least trying to look better, although consumer investigations such as those run by the Ethical Consumer organization strive to minimize greenwashing. As for employees, those of us who started the change and those of us who joined it are now finding opportunities to ensure that products are made in a way that benefits, or at least does less damage, to the environment we all live in. This is why I feel the need to be 'loud' about this: to support other people who see work in climate as essential, just as I have been supported. Ultimately, to be able to design consistently sustainable products, we need a tipping point so that such determination is a majority view.

I believe this is not unique to my situation, although in engineering we are very explicitly trying to create the future. Looking at surveys of public attitudes,[8] the majority of our communities are already concerned about climate, about

7 It's important not to forget that it's also a biodiversity job, as well as a job for a fair transition.

8 A government survey in 2022 said that 'Around three in four adults (74 per cent) reported feeling (very or somewhat) worried about climate change; the latest estimate is similar compared with the percentage who said they felt worried (75 per cent) around a year ago.'

biodiversity loss or about a lack of justice. As individuals, we feel disempowered and voiceless, just like me in that interview in 2021. Whether you work in engineering, in insurance, in health, in education, in retail, in your own home ... it's no longer likely that you are alone if you are concerned about what is happening. I now believe that the points of commonality we share with others provide opportunities to build communities for support and inspiration. In fact, there is probably already a group out there for people like you, and if there isn't, there is likely to be someone else who would join your group if you start it.

Every job, hobby or interest is a climate job.

WHY I TALK CLIMATE TO EVERYONE

Laura Thomas-Walters, PhD

Laura is Deputy Director of Experimental Research for the Yale Program on Climate Change Communication. Her research focuses on changing behaviours to help solve environmental problems, and she works closely with governments and NGOs to apply her results. She has both a professional and personal interest in climate activism, and has conducted a lot of research into climate activism as part of the Data Analysis and Insights Circle in Extinction Rebellion UK.

I study changes in behaviour regarding environmental issues. This means I try to figure out how to get people to behave in ways that are better for the environment. I've looked at topics from the illegal wildlife trade (persuading people to stop buying things like rhino horn or elephant ivory) to the impact of mobile-phone games on donations to help save the purple frog.

I like to think I'm pretty good at putting my money where

my mouth is. I'm a climate activist, I never learned to drive, and I'm trying my best to fight my own rampant consumerism tendencies. However, it wasn't until I started studying something called 'relational organizing' that I realized there was a big area of climate action I was completely neglecting. Weirdly, it's the simplest action, too – just talking about it!

Relational organizing is just people talking to their friends and family and encouraging them to do better. That might mean voting in elections, buying an electric vehicle, or flying less. It rests on the idea that we trust our friends and family more than random advocates from charities or the government, so they are also the best messengers to promote good attitudes and behaviours. If my friend recommends a new restaurant to me, I'm more likely to trust her review than that of a stranger online. If my dad wants to discuss the crisis in Gaza, I will try harder to understand his position because I know he is a good person.

My new research involved trying to persuade people who were vegetarians or vegans to promote plant-based diets to their friends and family. A plant-based diet (or even just eating less red meat) is miles better for the environment. This is something I know very well – I've been vegan for over fifteen years. However, I've lived in fear of being seen as the 'preachy vegan'. My heart sinks when people bring up diets because I really don't want to be seen as judging others. But I also felt like I'd be hypocritical if I tried to persuade other people to do what I wasn't willing to do myself – that is, just talk about it.

I decided to start with my mum. I knew she already understood the need to eat less red meat, but that she just didn't have enough motivation to make the change. So I asked if we could have a serious conversation, and asked her to think about the sort of world the children in our family were going to grow up in. I knew this would matter more to her than the health or animal welfare arguments. Happily, she's now stopped eating red meat for nearly a year!

I went on to talk to other family members, tailoring the conversation to the things that mattered to them. I also started talking about the climate more widely, and to strangers. I'd chat about the weather with my hairdresser or Sainsbury's delivery driver, linking it to climate change. To the bus driver I explained why I was wearing an Extinction Rebellion badge, and why I thought climate activism is important. When I overheard the plumber fixing my boiler telling his apprentice climate change isn't real, I forced myself to interject (and got a cheer from his apprentice)!

Doing this, I realized something – people care about the environment way more than I thought. The media likes to paint climate as a dirty word, and presents a view that most people are apathetic. It's just not true. Research, and my own experiences, show that actually most of us do support climate action, we just all think other people don't care as much as us! When I started talking about climate to everyone, I found that generally people recognized there's a big problem, and believed that the government ought to be doing more to protect the environment. They just weren't sure what *they* could do to help. Luckily, as an environmental behaviour scientist, there was plenty I could suggest.

Still, the number one easy thing to do is talk! Change the way society perceives climate change. Don't let the media and government dismiss it as a fringe issue. Share your experiences and encourage your friends and family to act now.

STAND-UP FOR CLIMATE: LAUGHTER MIGHT BE THE BEST MEDICINE

Tristram Wyatt, PhD

Tristram did his undergraduate studies and PhD in zoology at Cambridge University. Before retiring, he lectured in biology at Oxford University's Department for Continuing Education. The second edition of his book Pheromones and Animal Behavior *won the Royal Society of Biology's prize for the Best Postgraduate Textbook in 2014. He published* Animal behaviour: A very short introduction *with Oxford University Press in 2017. He's now an emeritus fellow of Kellogg College, Oxford, and honorary research fellow at University College London. Tristram joined Extinction Rebellion in 2020 and has participated in numerous actions with Scientists for XR. There's a recording of his very first stand-up at tinyurl.com/Wyatt-standup*

Suddenly I was next. I walked onto the stage, into the spotlight, picked up the mic, looked out into the darkness of the packed theatre and started: 'My name is Tristram and I'm a scientist. I'm an activist and my partner is not. He's tall and handsome, and so am I!' (I'm not tall, the rest is a matter of opinion.) Ten minutes later, after laughter and applause, my first try at stand-up comedy was complete. To my surprise, I had made an audience laugh about the planetary crisis but in ways that seemed to resonate, bringing them with me.

Eight weeks earlier, an email had been sent to Oxford University biologists offering a short course in comedy or poetry for scientists and engineers interested in communicating about the climate and ecological emergency. As a scientist turned activist, I had tried everything I could think of to wake people up to the climate crisis, and none of it had properly cut through. Perhaps stand-up would. In a moment of light-headedness, I signed up, choosing stand-up comedy as I thought it would be less frightening than poetry. When I confessed to my husband what I had done, he was so shocked he had to sit down. It was not what he expected of me.

I also didn't know what to expect. Each week, on Zoom, our tutor took us through brainstorming ideas that could be developed. The first exercise was to think of things that got us worked up and annoyed. Riffing on those gave us the first two minutes of what would become a ten-minute stand-up set. Each week we would perform the latest version to our fellow students and get feedback from them and our tutor. The most important feedback was simply, did it make them laugh? Our tutor, an experienced stand-up comedian, would also give us one-to-one feedback. The task each week was to use the feedback to rework our sets and add another two minutes. The rules for creating the comedy were that it should always punch up or be about ourselves, never punch down, and that it should be compassionate, that is, with a good heart.

The course culminated in a half-day workshop in Oxford where we met each other for the first time, and we could learn some stagecraft for live performance. The first lesson was practising picking up the mic and very deliberately putting the mic-stand behind us, in a ritual to calm ourselves after coming on stage. The next evening all of us performed in front of a full house of friends and family numbering some two hundred. The poets went first and were fabulous. I almost wished I'd tried the poetry course after all. Stand-up came after the interval, but our tutor said no wine until after we were all done. Stand-up is best done sober.

As I walked on stage, I was worried about forgetting crucial lines. We'd learned on the course about the difference between improv and stand-up. Improv is, as it says, improvised. Stand-up uses a carefully prepared script that works at its best because the audience is set up for each joke early on and the laughs come from either an unexpected twist or a planted shared reference. For stand-up, a missed line could jeopardize a joke later in the set because the audience has not been primed. The challenge is that for it to work you need to give it almost word-perfect, from memory, without notes.

Our tutor was the MC that night and she created a warm listening space by explaining that this was compassionate comedy, and the audience should encourage the performers by enthusiastic clapping whenever they might seem to forget their lines for a moment. That generous applause rescued me twice in my piece and gave me time to recall the next line.

Since that first terrifying evening, I've given my climate comedy set to a room full of office workers at lunchtime, when I had to ask them to imagine they were in a nightclub after a few glasses of wine. I've done it as an after-dinner speech at an Oxford University college, where I had to be my own warm-up act to get them in the mood. I've also used it at a Just Stop Oil fundraiser, adding that 'Just Stop Oil is listening to the science, and the government isn't.' It also lifted my talk, encouraging scientists to become climate activists, during the annual Darwin's Birthday Debate at the Natural History Museum in London.

Like so many other scientists turned activists, I've been campaigning for decades, doing all the conventional things like marching with placards, writing to my MP, futilely signing endless petitions. I was there at many Extinction Rebellion actions from the very beginning in 2018, but as a bystander rather than full participant. Seeing the Scientists for XR in their white lab coats brought me on board. It coincided with me thinking about writing the next edition of my textbook on pheromones, which has become a go-to book for my field – the most satisfying recognition in my professional life as a scientist. As I researched the update, I realized that most of the wonderful animals I would be writing about would become extinct in the lifetime of the students reading the book. I resolved that rather than spending my time writing obituaries for the animals as they were extinguished, I'd get into action with the Scientists for XR. Many of these ideas found their way into my set.

Getting people to listen is key – comedy may be particularly effective for this. There's a rapidly growing literature about using comedy and satire to engage people's attention. Humour allows taboo subjects to be raised and dissected. For example, films like the dark comedy *Don't Look Up*, in which two astronomers try to warn the world about an approaching comet, can reach new audiences with a climate message.

My husband does not do activism. I've never been able to persuade him to come to a demonstration, let alone an occupation. He also thinks that I tend to get too earnest about climate (and other politics) in a way that puts people off – it certainly turns him off. He feels my evangelism can reek of pomposity and posturing, which is not the way to persuade, not the way to reach out to the unconverted. By contrast, he's enjoyed the way my stand-up seems to get my message across even while, or really because, people are laughing. He's pushing me to do more, to write more material for a second set. Comedy gave me a voice that I did not know I had. It might not be for you, but speaking about the climate crisis can be cathartic. So many conversations are waiting to happen, and finding your own voice is an important part of the process.

POWER FROM PEOPLE: HOW THE FORCE OF THE RIVER THAMES INSPIRED LOCAL PEOPLE TO HARNESS THEIR SKILLS FOR A SUSTAINABLE FUTURE

Sophie Paul, MSc

Sophie has worked in hydrogeology, IT, ethical governance, campaigning and community energy. Her drivers are human and animal rights, climate and biodiversity, community-building and social justice, and dancing. They are all so vast it's hard to know where to start. How about a supported transition to a plant-based food system and replacing fossil fuels? Overwhelmed as to how to contribute, she has tried working in all sectors, with community and activist groups bringing the most reward. Here's one personal example.

Walking over Caversham Weir, five minutes from Reading train station, brings you to a small building, with a bright mural on two sides and on another side the Climate Stripes[1] that Reading is famous for. On the fourth side of the building are two huge Archimedes screws, flicking water as they rotate in the power of the Thames, generating electricity. More than 800 local people brought this about. We wanted to make a direct, visible contribution to green energy. In the process, we brought people together.

A Community Power Project

I am walking over Caversham Weir, the River Thames in spate thundering below, flowing on to London and the North Sea. I feel and smell the power – I want to capture that natural energy. Surely someone else wants to, too? And they do: a friend directs me to meet Tony, an energetic mind voracious in creating sustainability projects across the town of Reading.

It turns out that an application for permission for a hydropower plant has already been submitted to the council's planning department through the hard work of local sustainability groups. Together, we set up Reading Hydro formally as a Community Benefit Society (CBS), a kind of co-op. We are a diverse collective of warm and tenacious characters, scaling the mountain of expertise, bureaucracy and administration needed. Many individuals at some point make a crucial contribution to keeping the project alive. Constant planning, networking and vigilance go into attracting the skills and characteristics necessary to make an effective and responsible organization that is, above all, rewarding to be a part of. No community organization can survive without this.

1 These simple stripes of blue and red show how global temperatures have risen over some 200 years.

It is hard work to develop community energy. Add in building in the middle of one of the largest rivers in the country, with its competing users and environmental considerations, and the permits and technical considerations escalate. As Chair of the CBS and knowing what potential Reading has, I'd like to help expand what this project covers: bringing in local strengths such as IT and open source data, business collaboration, unique street art, educational prowess, and regeneration of local nature. The spectrum of skills needed is broad, for which Reading proves the perfect town.

The project starts small, with some helpful grants for the planning application and early work. We raise funds for a detailed scheme design with a pioneer share offer, enthusiastically supported by local people who are prepared to take on the high investment risk at this early stage. From this we build our full business plan.

For two years we work hard to involve an exciting new technique to generate hydroelectricity. This new technology is based on the Venturi principle, which is to do with changing water pressure by restricting its flow, leading to improved power output. It would also simplify the engineering needed for low head (low weir height) situations such as ours. However, at the time, this technique proves too immature to obtain the regulatory permissions we need. So we return to tried and tested reverse Archimedes screw technology.

Obtaining myriad regulatory licences is our biggest hurdle, particularly with the many different departments of the Environment Agency (EA) that are engaged in protecting the river. Finally there's a breakthrough when the EA brings its relevant departments together for us under a single point of contact. This ensures that our permits come through in time for us to benefit from a government funding scheme that makes the project financially viable. A local business agrees in principle to buy our energy. This clinches our economic viability.

Money is tight in hydropower, given the high upfront capital costs. It's a long-term investment that tends to need government support, such as the feed-in tariff we have, or hopefully one day, a fair electricity selling mechanism via the National Grid for small generators.

We raise capital for the build with a hugely successful main share offer, with 750 mainly local people investing. We amass over 150 volunteers over the planning and build phases, some working on our project more than full time. Only the scheme designers, occasional specialist consultants, a project manager and contractors are paid. Many crucial skills are volunteered, including the vast amount of technical, practical, organizational and finance work needed.

When we are finally ready to build, the pandemic hits. Our interactive AGM of 120 people tests our urgently learned Zoom skills to the max just days into the first lockdown, with many complexities to explain and questions to field. Six weeks later, amid ongoing restrictions, a handful of us go out with a couple of spades and wheelbarrows to break ground before our planning permission times out.

As the pandemic weeks roll on, we learn to make the most of people having their calendars empty. We fill our lives with socially distanced river clearance and site preparation, and frequent video meetings. Sadness in my personal life is replaced by happiness at getting soaked during river clearance in the long heatwave, and the camaraderie of energetic people. Some team members are not so fortunate, as they can never come to the site due to shielding and health restrictions. We have to keep together online, with these people powering the project behind the scenes.

When the civil engineering contractors arrive, the wettest October on record rains down, flooding the river. Weeks of build time are lost, racking up huge costs. Two more well-presented share offers are successful in raising the extra funds. We cut costs by deciding we can build the turbine house and technical section of the fish pass ourselves, enabled by the unquenchable Tony using his own builders' insurance. Obtaining insurance can be tricky for young community organizations, particularly when there is physical work involved. The self-build element saves the project tens of thousands of pounds, yet demands even more of volunteers on site and behind the scenes.

Currently, if as a small generator you sell your electricity directly to the national grid, you receive a small payment per unit for it. Whereas if you can sell directly to a single customer, you can charge a mutually beneficial reasonable rate. For this, the project needs a connection by direct wire to the customer, who is ideally big enough to use all the energy produced, so that little continues on into the grid to be sold at the lower price. Cables large enough to cope with the output are unwieldy and must be buried in a regulated way. Every metre adds extra cost, but we can't take the direct route under the boating lock to our customer, as that is too risky for the EA. So we plot a route around this, under an open part of the river, and secure permissions from the council to connect our cable under their land to our customer.

We can't afford to contract out the job of pulling a 400-metre cable through the duct we've had specially drilled under the River Thames. So in the middle of the third lockdown, we muster ten teams of six volunteers, all socially distanced and trained online. Throughout the Friday night a relay of couples stay up to babysit our valuable cable on site. Together, the sixty cable-pull volunteers spend the whole Saturday getting that cable laid, buried and connected. The teams and a powered winch complete this in thirteen hours against the odds, with walkie-talkie handlers aiding coordination across the river. Someone even runs off to get tubs of vegetable margarine when we run out of the expensive natural lubricant that encourages the cable through the underground ducting. Even the extra people lending tools or bringing round home-bakes are crucial in making the day a success.

Another high point – alongside the regular heartaches and headaches – arrives when the two long, 2.6-metre diameter turbines (the Archimedes screws) finally trundle in on the back of trucks from the Netherlands. They are craned into place with a small select crowd cheering them in, for that's all that's allowed in the continuing pandemic. The turbines have been delayed at the factory by COVID-19. In the weeks to come, Tony and I are moved to tears to find that our fellow directors have decided that the turbines will be named after us.

That summer, the hydro scheme is commissioned at last. We have a party on site, still with COVID-19 restrictions, yet bringing together key people from the life of the project to celebrate. My younger daughter's band plays live, our new fish pass burbling between them and the party, my elder daughter videoing it all. Our local Member of Parliament performs the opening ceremony, with the event televised by both local BBC and ITV.

Can we all now rest? Oh, no! Now we have to learn how to operate and run a power plant on a shoestring. Again it's teams of volunteers training up and turning up, this time under a 24/7/365-rota. A drought delays our actual operational start by several weeks, giving us time to finesse our plans. Becoming Operations Director, my reward is in building a team, devising together how to operate and maintain the scheme as volunteers. Whether it's keeping the machinery going, maintaining the site or clearing debris off the grills protecting the turbines, we cover urgent action, routine work and in-person checks. We have much to learn and frequent urgent situations, as well as huge amounts of process-building. It feels like a continuous firefight. Reaching out to volunteer networks, Facebook groups and local specialist collectives, fresh recruits join familiar faces.

In parallel to the new Operations and Compliance Teams, we have morphed our build-phase teams into operational teams running Administration, Finance, Digital, Communications (including media and education), and People, with the chaired Board of Directors representing each strand. There are inevitable stresses between the front line and administration of the project.

Live interactive data streams to our website. This includes extra sensors via the open source Things Network. We are the first hydro project we know of to have our own multiple access operational login, coded by our team using open source software with support from the equipment suppliers. Thus the team of operations leads can offer support from anywhere – our homes or work. Nevertheless, we often have to go on site at unexpected hours, to check out something in person that the systems or locals have alerted us to.

Being both Chair and Operations Director is at times conflicted, so I'm keen on succession planning. Handover takes several months to come about, with high demands on the Chair at this transitional stage leading to two co-Chairs taking over. Work at all hours and signs of burnout continue for me, so I manage to hand on the Operations Director role several months later. I turn to a nomadic life to recuperate, knowing that this amazing project is in capable hands.

It took seven years after the planning application to get an operational community hydropower plant. The failure rate for hydro projects is high, yet Reading Hydro made it. All this work for 46kW of power production, generating the annual amount of electricity used by ninety average homes. Was it worth it? Did we make a difference? On the plus side, this renewable energy scheme may last for a century or more. It's become a scientific educational landmark in the heart of our town. As well as paying back local people who believed in us, electricity sales will build a community fund. This will support new local sustainability projects in decades to come.

Yet what I'm most proud of is the diverse and proactive community of people we brought together and the friendships made. People are needed to pass on the baton to keep generating in the decades to come, raising funds for the increasing need for sustainability and local solidarity. Anyone building a community energy scheme has to put in place continuous succession planning to ensure that the years of focused hard work that made it happen keep on giving. All of us, for whatever length or depth of involvement, become more than who we are when we work together.

So is a community energy project enough to stop the climate and ecological emergency? Of course not. I see short-termism and lack of vision in governments, corporations and mainstream media. Their entrenched self-serving systems are in the way of urgent progress. That's why I've moved on to join the campaign groups Scientists for Extinction Rebellion and Animal Rising, to push for well-informed, urgent change. Nevertheless, I feel that whatever comes to pass, resilient and resourceful community-building is part of a possibly survivable future. This project and the people in it give me hope that it can be done.

PART 3

SCIENTISTS IN PROTEST

'A STORY OF STATUES AND SANDWICHES'

HOW I FOUND MY COURAGE

Aaron Thierry, PhD

Aaron's fascination with the natural world developed when he was growing up exploring the rugged hills and valleys of Gwynedd, Wales. This curiosity led him to study zoology at university, then do a PhD in ecology, and from there to working as a research scientist on a project investigating rapidly thawing permafrost in the far north of Canada. His studies led him to become acutely aware of the harm currently being done to the Earth's life support systems. With such knowledge comes responsibility and Aaron has become a passionate advocate for environmental and social justice, and has since shifted his research field to focus on studying the growing climate movement. He has contributed to numerous successful campaigns, including those fighting for fossil fuel divestment and clean air zones. However, as the planetary crisis worsens, he has felt the need to go further and has started taking part in acts of civil disobedience focused on the need for urgent climate action.

Anxiety had crept its way through my body. I'd slept in fits, full of doubts, having gone to bed the night before still unsure as to my decision. Now the day had arrived. I had risen at dawn and was gazing across the rooftops of London from an attic bathroom window, chimney pots glowing in the mauve light. In that moment of calm, I was forced to make my choice. Was I the person I hoped I was? I cast my mind back to the lessons learned while exploring the glacial valleys of my youth, the years of reading and writing scientific papers, my encounter with the front line of the climate crisis in the Arctic, and then I let myself cast my thoughts forward to imagine a future conversation, perhaps with a younger relative, asking me to account for what I'd done with the knowledge I had gathered. And with that I knew. I turned to get ready for the long day ahead, but as I was about to brush my teeth, I caught my eye in the mirror and a wave of relief pulsed through me. I was that person, and suddenly tears welled in the eyes of my reflection. I had found my courage: I really was going to superglue myself to the Department of Business, Energy and Industrial Strategy.

I hopped onto the Tube and wove my way to the secret rendezvous – a café near Whitehall. It felt ridiculous to be sitting there miming drinking coffee (I was trying not to fill my bladder given what we were about to do) while also not acknowledging my incognito collaborators sitting at every other table. Even more stress-inducing was the police car parked on the street right outside. At the pre-arranged time, we left in our ones and twos. On paper it had seemed like a good idea, but in practice the room went from overflowing to half empty in a matter of minutes – surely that would arouse suspicion? The paranoia was really kicking in now. Having made my decision, I really didn't want all our plans to be for nothing. I shuffled past the police car and down a side alley, trying my best to not attract any attention. As soon as we were all out of sight of the police car, we divvied up our supplies. We each put on our white lab coat (symbolizing our scientific credentials), took some posters and a pot of wallpaper paste, and pocketed the all-important tube of superglue!

This was it; the moment had arrived. My adrenaline started to spike. We hurried into position along the glass-fronted building in the heart of Westminster. Leading the charge were the banner holders – their flag simply read 'End Fossil Fuels' in bold, bright letters. So far, so good. Behind them we'd split into pairs. My partner wiped generous dollops of the wallpaper paste to the glass, while I unfurled my set of posters and stuck them soundly to the windows. Each poster was the first page of a scientific paper from a leading academic journal conveying the urgent need to halt fossil fuels to prevent climate breakdown. Mine was titled 'Committed emissions from existing energy infrastructure jeopardize 1.5 °C climate target', and had been published in *Nature* a couple of years beforehand. I could see several security guards speaking into walkie-talkies, but they weren't trying to stop us. Next, the glue. I pulled off the stopper and applied superglue liberally to the palm of my left hand, making sure to also squeeze a line of the sticky substance onto each of my fingers. Then with a deep breath and a surreal sense of disbelief at what was happening, my fellow scientists and I stuck ourselves to the government building.

'NEW OIL AND GAS EQUALS DEATH,' I bellowed, quoting the patches carefully pinned to each of our lab coats. I went on shouting as loudly as I could: 'The UN Secretary General has said that any governments that are continuing to invest in the production of new fossil fuels are "dangerous radicals". Our government are acting like dangerous radicals!' The shouting helped release some of the tension that had built within me, and I started to relax slightly and reflect on what we had just accomplished. We were taking this action just a few days after an IPCC report had been published,[1] detailing that the majority

1 This specific IPCC report was 'Climate Change 2022: Mitigation of Climate Change', the Working Group III contribution to the Sixth Assessment Report (AR6).

of fossil fuels already discovered must remain unburned if we were to meet the carbon reduction targets that all the world's governments signed up to in 2015. The science, as set out by the IPCC, was absolutely clear that if we missed these targets, large swathes of humanity would experience truly catastrophic impacts. Yet, heedless of this plain warning repeated (yet again) by the scientific community, our government had announced that it was to issue licences for new oil and gas projects in the North Sea! We were therefore determined to leave an unambiguous message that Kwasi Kwarteng, who was at that time the Minister for Business, Energy and Industrial Strategy, was ignoring the scientific evidence. If they wouldn't read the reports we would paste them to the windows, so that the public would hear about it. After all, we reasoned, actions speak louder than words. We had glued ourselves to the building in this act of civil disobedience to show that we stood firmly by the science.

Within minutes the police were on the scene. As someone who normally avoids any drama or conflict, my heart began to race again. I turned and caught the eye of one of the other scientists glued to my right – she was standing next to a blown-up page titled 'Assessing "Dangerous Climate Change": Required Reduction of Carbon Emissions to Protect Young People, Future Generations and Nature.' We smiled nervously at each other, unsure as to what was about to happen. Then I caught a look of determined resolve on her face, she nodded at me with satisfaction, and I nodded back, heartened. An officer came over to me to ask what we were doing. She was about my age and wore a yellow, high-vis jacket and a stern expression. I explained that we were protesting government inaction on the climate crisis. Next, she asked about how long we intended to be there. I explained that we were glued on and had no intention of leaving. She then asked why I was dressed in a lab coat, and I explained that we were all scientists. That startled her. 'Really? What, you're actually scientists?' 'That's right,' I replied. 'I'm an ecologist. I've worked up in the Arctic studying carbon cycle feedbacks from the massive increase in wildfires in the boreal forest as a result of global heating and I'm terrified by what I've learned. The politicians in this building are ignoring our warnings and continuing to double down on fossil fuel investment. It puts us all in danger.'

After talking for a while longer, the officer wandered off to confer with her colleagues about what to do with us. Members of the public gathered nearby, trying to figure out what was going on. A few heckled us: 'What do you think you're going to achieve by doing that?' But most called out 'Thank you!' or 'Well done!' Journalists showed up and started taking photos and interviewing us. I was relieved that the press release we put out appeared to have been picked up, and I hoped that it meant our message would be in the news that night. There was a lot of hustle and bustle around us – videographers, live-streamers, social media posters trying their best to get our story out on Twitter or Instagram, and time started to whirr by. I was becoming thirsty but I was still determined not to drink anything as I had no idea how long we'd be stuck to the window. My arms started to cramp, so I tried with some difficulty to move position, telling myself that at least I could feel my fingers. I realized that I hadn't given any thought to how I would attach myself, and I'd ended up doing it in the most awkward way; I must have looked like I was attempting some weird yoga pose with my arm outstretched behind my back. I chuckled at myself and the absurdity of the situation. But the humour of it wore off quickly and I started feeling dejected that it had come to this: that reason and evidence alone had failed to persuade governments to act, that our elected representatives had succumbed to the power of the vested interests of the fossil fuel industry, that scientists like myself felt we had no better option left than to take these risks in a desperate effort to signal our level of concern and desperation. At just that moment a journalist pointed a camera in my face and asked me to account for why I was there. I blurted out 'The government's insane, and I don't know what to do, other than to do this, to try and get the attention that we need to wake the public up!'

A few more hours went by, and we were told that we were to be arrested. First a specialist police team needed to detach us from the glass. A chemical solvent was squirted around my fingers, and each started to peel off easily. I thought being removed might be painful and leave behind a few layers of skin, but the officer worked very carefully and within a few minutes I was free, with no glue residue on either me or the window. My arresting officer was Romanian, and only a couple of years on the job; he read me my rights with a heavy accent. In the van on the way to the station we talked about my motivations. I asked him to look out at the River Thames and for him to imagine how much of London would be lost if, as my colleagues warn is increasingly likely, large parts of the ice sheets in Greenland and Antarctica melt. I asked him to imagine the chaos of the city the first time the flood defences are breached, and homes flooded. I asked him how many times he thought we'd try to rebuild. Next, I recounted how we know that the Earth system can change dramatically from one state to another; that I grew up in a glacial valley in Snowdonia in North Wales, which 20,000 years ago was a barren wasteland under half a mile of ice, that the average temperature of the planet was just 4.5°C cooler back then. I explained that we're currently on course for human pollution from fossil fuel burning to warm the planet by over 3°C by the end of this century – a huge change in just a few decades, and one that would push vast areas of the planet beyond the tolerance of human survivability. Billions of people would be displaced. He nodded, and said that he'd heard such warnings before, but what would gluing myself to a window do in the face of such a challenge? I said it was an act of faith that humans could change course. He explained that he'd grown up under a dictatorship, and he'd seen first-hand that if you try to go against the system, the system would crush you. Better by far, he suggested, to play by the rules and try to gain influence to change the system from within. 'But ...' I asked, 'wasn't it ultimately a popular rebellion that forced Ceaușescu from office?'[2]

We were charged with criminal damage and our trials took place in the months that followed. The nine scientists who were arrested that day were all eventually found not guilty, on the grounds that we were peacefully exercising our right to political expression. It was a glorious reminder that we must never take our hard-won civil liberties for granted. We were supported in defending ourselves in court by a team of dedicated lawyers, funded in large part by a humblingly successful crowdfunding campaign that led to generous donations from the public. Reflecting on what we did, I'm proud of our action, though I recognize that around the world many activists confront far greater risks on a daily basis. Over the past decade, the international campaign group Global Witness has recorded the murder of an environmental defender almost every other day. I contemplate such sacrifices regularly, and have immense admiration for those who give so much to protect the Earth. Nevertheless, for me at least, deliberately choosing to publicly break the law was a deeply uncomfortable experience, and something I couldn't have done alone.

I gained the courage I needed from the knowledge that I was acting with other brave scientists who I intensely admired, for we all trusted one another and knew we would stick together no matter what. I took courage from the actions of the youth strikers around the world who had been acting on the scientific warnings and in their millions had been missing school to fight for their futures. I found my courage in recognizing I was just

2 The dictator Nichloae Ceaușescu (1918-1989) was overthrown in the Romanian Revolution.

one tiny part of a whole, stretching back centuries into the past, and forward centuries into the future, fighting for a more just and loving world. I dearly hope that our taking part in this action will give some small courage to others to act, and that they in turn will inspire others. For I'm convinced that the safe and just future we all so desperately want will only be brought about if far more of us are able to find the courage to come together and rebel for life.

A STORY OF STATUES AND SANDWICHES

Abi Perrin, PhD

Abi was following a fairly traditional path in academic research, studying at the University of Cambridge and undertaking postdoctoral work at medical research institutes, until the enormity of the climate and ecological emergency began to feel overwhelming. Around this time student climate strikes were going global and Extinction Rebellion protesters were filling the streets of London – the city where Abi was living, at that time working as a malaria researcher. Having tentatively joined those protestors and soon finding the other scientists in their midst, she has increasingly embraced the idea that scientists have a role to play in pushing for systemic change. This has involved realigning her priorities and work to focus much more directly on climate and environmental communication and action, alongside being an active member of the scientist–activist community.

Some context ...

These words are based on those I wrote in reflection shortly after the Scientists for Extinction Rebellion action at the Department for Business, Energy and Industrial Strategy (BEIS) in April 2022. The protest was prompted by the UK government's strategy to maximize the extraction of North Sea fossil fuels.[1] Nine scientists were arrested at that protest; I was one of them and while it wasn't the first time I had taken similar risks, it was my first arrest. Snippets from that day and those that followed remain very vivid to me. Those experiences and events have had lasting impacts on my personal relationships and my interactions with colleagues and institutions in the course of my work, as well as on how I view the roles of protest, policing and the courts in our society. Knowing that, it feels quite strange to realize that the memories that play repeatedly in my head aren't in any way dramatic; they're predominantly the small, human moments of support, connection, compassion and humour. I still find myself hearing the gentle drumming and singing of one of our supporters, feeling the warmth of the relative strangers who made it their job to take care of the arrestees, replaying a conversation between three scientists in the back of a police van, and reliving a debacle over a hummus and chipotle wrap.

The story ...

It's a place I never expected to be in: sitting in silence in the back of a police van with two other scientists. Neither (I assume) was it

1 The chapters 'How I found my courage' by Aaron Thierry (page 132) and 'Unstuck in time' by Lucy Hogarth (page 146) are different perspectives from different moments of this same protest action. The rationale behind our actions that day is described in Aaron's account.

a situation the arresting officers had anticipated, as they escorted the three lab coat-wearing detainees and observed the (valid but mathematically questionable) slogan 'new oil & gas = death' that we were adorned with. I'm not sure I can quite describe my feelings at that moment – maybe tired, possibly sad, strangely calm but at the same time not at peace. I can remember some of the thoughts swimming around in my head. How did nine scientists get to this surreal place? How had so many years of scientists' warnings failed to generate political urgency on climate action? How had it got to the point where each of us saw arrest as a worthwhile risk in our own efforts to amplify these warnings? Why was it that these important scientific messages weren't *already* consuming the consciousnesses of powerful people whose strategy and priorities we were trying to challenge via our protest that day? Had we made a difference? Were all the other scientists OK? Did I say the right things to those journalists? What will my parents think??

Somebody broke the silence as the van made a loop around Parliament Square. That scientist, Stuart,[2] started chatting animatedly about the statues we were passing. At least three of the figures commemorated are there because of their roles in what we now see as powerful and influential protest movements – these are people who we remember because their actions contributed to societal change that protected and liberated others. I don't equate our protest that day as anything close to those of people who experienced huge oppression, significant violence and great sacrifices. However, being made aware of their presence as we made that journey was a grounding and much-needed reminder of what we can achieve if we stand together, challenge harm and injustice, and demand or create better alternatives.

2 You can read about social psychologist Stuart's perspective on the climate and ecological emergency in his chapter 'Humans: the cause of and solution to all the environment's problems,' (page 20).

None of those figures in Parliament Square made the changes that we now celebrate happen by themselves; they were part of mass movements, full of people who had no guarantee their actions would be effective or whether the risks they took would feel worthwhile. Those movements probably contained many people who worried about how they might be portrayed, whether they'd said and done the 'right' things at the right moments, or what their parents might think. But those people knew they had to try, and I am glad they did.

Stuart continues talking about the history and effectiveness of social movements that use nonviolent civil disobedience, and Emma[3] – the other scientist in the van – starts joining in.

'No talking in the van,' says one of the police officers.

After a few minutes, we arrive at our destination, Charing Cross Police Station. The three of us are then separated and won't see one another again until we are charged the next day. We are each taken through a standard procedure to process us as 'prisoners', but my arrival is complicated by the presence of a sandwich in the pocket of the lab coat I am wearing.

The four police officers involved in the process can't agree about the correct fate of the sandwich. I'm told that the sandwich policies are not consistent between London police stations. I'm aware of their discussion about the relative merits of the approaches that would be taken at each of their respective stations (past and present) in such circumstances. The custody sergeant interjects to inform them that – due to the detainment of more environmental activists than anticipated – the station currently has no vegan or vegetarian food left, which might be a reason in favour of sparing the sandwich in question from destruction. I'm asked to provide assurance that the sandwich

3 You can read part of Emma's story in her chapter 'Sentenced for life: why I went to prison for nature' (page 150).

derives from a reputable retailer and that I have not tampered with the sandwich. This issue is not resolved before I'm taken to the cell where I'll spend the next twenty-four hours.

For those hours I'm back where I started this story – with those questions swirling around, and flashes of the day and what led up to it punctuating them. But now there are no friendly scientists present to lift the mood and those all-too-familiar feelings that ultimately led me here return. A pervasive sense of grief for all that's already lost through our collective failure to listen and respond to the warnings from science and the communities on the front lines of climate catastrophes. 'The Dread' that dominates when I can't stop envisioning possible versions of the future we are hurtling towards. An indignant frustration that playing by 'rules' that I'd internalized (work hard, trust authority, don't make a fuss ...) hadn't – and wouldn't – give us a fighting chance of protecting the beautiful, rich diversity of people and species alive today from the devastating impacts of climate and ecological breakdown. A guilt that, with all the privileges of my upbringing, education and career, it took me the time it did to recognize the global harms caused by my own lifestyle and to realize I had a responsibility to act. That feeling is complicated by the knowledge that my actions to try to address that big picture will also cause discomfort, concern and annoyance in people I love. I sit with a fear that, at a time where we desperately need to be pulling together, we are instead seeing further polarization and division in our societies. And I ponder with an exhausted rage how it's even possible to detach from the huge injustices of the world we live in and, instead, occupy ourselves with comparatively trivial concerns like whether sticking a poster to a window is a criminal offence or how to ensure the proper implementation of a non-existent sandwich policy.

It's hard to know how to end this story. Perhaps that's because it's not yet over. Our actions that day did not change the government's energy strategy, but we are still seeing ripples that give me some hope that they played a part in spreading important messages, growing and strengthening the bonds within our 'activist' family, and rallying more scientists to step out of their labs and onto the streets in protest.

UNSTUCK IN TIME

Lucy Hogarth, PhD

Lucy, author of the chapter 'Saying goodbye to the Universe' (see page 98) describes her experience of one climate protest from her perspective as an autistic astrophysicist.

My favourite author of all time has to be Kurt Vonnegut. His writing is always where my head goes when I think about what it's like to be autistic. His narrative style is like hopping between different key moments in a character's life; jumbled up and only truly comprehensible once you have finished the story. That's often how I experience life. Perhaps it's the low threshold autistic people have for PTSD, but I can find myself drifting from one moment to another sometimes, pulled in different directions when some synapse fires after hearing a familiar sound or smell or just noticing a familiar feeling. One place I keep returning to is 13 April 2022, outside the Department for Business, Energy and Industrial Strategy (BEIS), where twenty-five scientists, including myself, stood in protest of government inaction in the wake of the Intergovernmental Panel on Climate Change's report. I don't return to the whole day, just moments of it, disordered and, like Vonnegut's stories, only truly understandable once you have seen all the moments together.

I'm not new to protesting about the climate crisis; it's been years since I joined Scientists for Extinction Rebellion. That hasn't made me a 'hardened' protestor, though. Each protest, each action, costs something of me and I become a little more unstuck in time. Protests are not easy places to be, even for a neurotypical person, but for me, they're moments I know I will be dragged back to whenever I start to feel anxious, scared or unsafe. This is the first time I've sat down and written an account of one of these unstuck moments and I can feel myself being drawn away as I type. This is not a moment-by-moment account of what happened on 13 April 2022. I won't mention protestors' names here, but instead I will recall how I experienced it and continue to experience it, over and over and over again. It starts with someone grabbing my hand in my coat pocket in a café near the BEIS building.

It's her walking past me and squeezing my hand tightly in my pocket. We're leaving and it's time to get ready, time to hide around a corner and put our lab coats on under our jackets. I stuff as many posters as I can into my jacket and I feel deeply conscious of the uncomfortable sensation of paper under my arms. People are talking and buzzing, and I try to remember to listen, to focus on the immediate. Remember to smile, remember to make eye contact, remember you're here.

The next moment surges quickly now. We're running towards the BEIS building together and I'm trying not to let my posters fall through my coat. Then it's time to stop, time to paste the posters onto the glass windows. I turn around to my left and see scientists bringing out their tiny pots of superglue. My stomach turns at this moment, not only through fear for them, but also through shame that I am not among those with a tiny superglue pot; I'm not glueing onto the building today. We stand a few feet in front of the gluers and wait. My guilt waits with me and keeps turning my head to those behind me. A lot of disobedience involves waiting with regrets.

I have to keep listening to what they're saying. People are talking and I keep drifting. Drifting back to a hand squeezing mine, the finality of it. Keep listening, keep talking, make eye contact. I can't drift away. I'm here.

It changes when he runs at one of the windows with a spray can in his hand. Sometimes I see him leap, or maybe he flew! He sprays a huge Extinction Rebellion logo onto the window and then he's pulled back, pulled back somewhere. I don't see him. There's yelling and movement all around me. He can't have been pulled to the ground. He just disappears and I can't see where the police took him. We've got to be still though, keep the line, but I can't see him and I can't breathe. Then she sprays the window, but she is swarmed by more police bodies. Her wrist twists as they push her towards the ground and I hear a crack, but it's just in my mind, I think. Her can sprays out in front of me as the police grasp her free hand and the blue mist slowly falls to the ground. The smell is acrid. I hear my friends shouting, 'She is glued on! She is glued on!' Don't the police know? Are they going to twist her off the wall? Don't hurt her. Why would you hurt her? I can still hear her wrist cracking, but the noise is not real.

The police bodies multiply and some of them have faces, by which I mean I notice their faces. The police with and without faces line up behind all of us holding our signs, separating us from the gluers. I look behind me, through the gap, between two police with faces, to her. Her eyes are red and they flicker behind resisted tears. Her hand is on mine in the café again, squeezing mine in reassurance. Her tears rise through my throat and out through my eyes. I can't restrain them, not when I see her holding them back. How dare they do this to her eyes. A policewoman with a face looks at me with a sad curiosity. My tears puzzle her; she doesn't know I'm crying the tears that my friends are trying not to cry.

Finally, I can see him, half in darkness in the police van. He's staring out as she is carried out by a new swarm of officers. I don't remember their faces. Behind me, her eyes are still red. A policewoman with a face tells me to move aside so they can take her into the van with him and her. This moment takes longer than the time it occupied. I want to turn to face the policewoman, with my eyes and cheeks drenched by the tears held back by my friends. Sometimes, I do turn around and block their path, but I remember that she made a choice to be arrested, and it's a choice I have to respect. I move aside and she passes with the policewoman. She's taken into the darkness where he is staring out at me. They're all taken from behind the wall of police with and without faces. She walks, he is carried, she is carried, I can't remember the rest. My tears leave me too quickly and thickly to see. I keep saying, 'I just want them to be OK,' and I have forgotten to listen, to make eye contact, not to drift away. The tears ground me here though. I am unstuck in this moment and it will keep drawing me back whenever I forget to stay still. It's too much here; I feel too much here.

The cost of taking action is high, for any person. The consequences of these moments are still being enacted for the nine scientists, my nine friends, who were arrested. It's a price they decided to pay to try to disrupt a species-wide indifference to its own future pain and suffering. These moments are the price I paid. Paying in these moments, though, costs less to me than waiting as the chances where I could have done something to stop this globally scaled madness travel past me. Every second, every fraction of second, is an action not done, a choice to do nothing, an opportunity passing by and lost. I can live with being unstuck in time. They are moments in which I chose to act and I have to believe that those choices are worth it. The alternative is to be firmly stuck in time with the knowledge and shame that I let those moments pass me by.

I am unstuck in time, but I have no choice.

SENTENCED FOR LIFE: WHY I WENT TO PRISON FOR NATURE

Emma Smart, BSc

Emma was enthralled by the natural world from a very young age. At fourteen, she told her careers advisor she 'wanted to save fish' and went on to obtain a degree in marine biology followed by PhD research on Mexican and Arabian freshwater fish. She discovered a new species of fish in Oman, advised local government, worked with the International Union for Conservation of Nature and established her own species and habitat conservation project. After several years working with the World Wide Fund for Nature (WWF), her belief that her work within academia and NGOs could meaningfully contribute to action on the escalating biodiversity and climate crises wavered. Emma transitioned to activism, starting with direct action against fox-hunting and the badger cull in England. She joined Extinction Rebellion in 2019 and in the five years that followed devoted herself full-time to direct action. She has supported activism with Scientists for Extinction Rebellion, Animal Rising, Insulate Britain and Just Stop Oil, and co-founded the rewilding group, Wild Card. Emma has been arrested seventeen times, and spent two months in prison, including twenty-six days on hunger strike. She has also been held on remand in prison for fourteen days and continues to face multiple fines and impending court cases.

To be a scientist is to live a lifetime of wonder and joy. Science is not just about observation, experimentation, discussion and conclusion. It is not merely a job. It's an existence of challenging, analysing and striving for truth. It's an opening of the eyes and mind to everything that's around you. Scientists are constantly asking more questions than they will ever find answers for – which is what makes science so utterly compelling and awe-inspiring. For me, to be a scientist was an opportunity to stand up for the millions of species on the planet. In the field of wildlife conservation, I believe that the fundamental objective of our research is to find practical applications that benefit the species and habitats being studied. Scientists of the natural world have a duty and responsibility to take a stand for the life they study.

During my time in academia I often felt frustrated and isolated. For me, gaining knowledge simply wasn't enough if that unique wisdom and understanding was not then applied to positive change for our natural world. As a conservation ecologist I always felt that a motivation to preserve should be an inherent necessity alongside research within this field.

As the biodiversity and climate crises intensified, the joy I gained from science turned into increasing pain, anger, frustration and grief. I felt scientists were no longer working towards conserving the natural world, but instead were documenting its demise. I no longer wanted to be part of identifying the problems; I wanted to be part of instigating the solutions.

I naively assumed that once science was in the hands, eyes and ears of decision-makers, they would act on that information. I was wrong.

For years I stood in classrooms, lecture theatres and laboratories, in rivers, ponds and streams, in NGO offices, government boardrooms, corporate meeting rooms and conference halls. That felt right. I thought at the time, 'This is where I'm supposed to be.'

When the climate movement began gaining momentum in 2019, I realized that the place I needed to stand was on the streets. To take a real stand during times of crises you need to take truth to power, not publications. That power lay in our governments, industry, media and financial institutions.

Two years later, I am standing on the side of a motorway. I take a deep breath and step out into the slow lane, leading a team of fourteen people onto the M25 – the ring road around London and one of Europe's busiest motorways – at rush hour. Traffic slows and stops. Banners are held by our team, who are all wearing orange high-vis. I take out a small tube of superglue, squeeze it across my hand and glue myself to the cold tarmac.

Then I'm in handcuffs, in the back of a police van, at a custody desk, a fingerprint machine, a cell. Two months later I'm standing in court and address a judge:

'The prosecution is here today by her own words to "ensure future compliance". Well we are here to "ensure future survival".

'I am proud of my actions and I stand by what I did.

'I have worked for almost twenty years in wildlife conservation; I am a scientist and a prison sentence will take me away from valuable environmental work that I am involved in.

'I used to believe that my place in fighting this battle, and it is a battle, was in a university, in a research lab, in the field and at my desk. I now know that where I stand right now and where I'm likely to go is the place where that fight must take place.

'I believe that my intentions are morally right, even if my actions are deemed legally wrong.

'This court may see me as being on the wrong side of the law, but in my heart I know I am on the right side of history.

'I will not be a bystander.'

I find myself in the back of a prison van, a cramped, white plastic cubicle with a small square window to the outside world that is scrolling past. The crowded streets of shoppers, a drive-by of consumerism and greed, designer shops – a tiny window view of a huge systemic problem. A long queue of people waiting to buy fast food from a burger chain. My heart is heavy but my mind is clear. I make an enraged, yet composed, decision to stop eating. They could take my freedom, but they couldn't take my commitment to the cause. A hunger strike would be a continuation of my protest demanding action on climate breakdown, a nonviolent escalation to demonstrate how serious I am, the sacrifices I am prepared to make.

After two and a half years of nonviolent direct action I am standing in a prison cell. Bright white courtyard lights burn through the cell window, creating sharp shadows of bars on the wall next to my bed. It is 2 a.m. and completely silent. I wake several times during the first night, each time a stark realization of where I am, like a nightmare you don't leave on waking. I peer through the window hoping I'd be able to see the sunrise. There are five narrow sections of scratched glass between four thickly painted blue metal bars. As it gets light, I become desperate to see or hear birds. A buzzard calls – I can't see it but I cry for the first time.

I continued my hunger strike for twenty-six days – one day for every failed COP meeting since the UN Framework Convention on Climate Change was signed in 1992. I found focus and control in a situation that was orchestrated to silence and stifle my freedom to protest. My limbs were narrowing, my skin flaking and my head dizzy if I attempted more than a slow walk along the corridor, but I felt stronger and more focused in that moment.

My method of protest was extreme, but these are extreme times. A draconian clampdown on nonviolent protest combined with the radical love and intense urgency of climate activists will inevitably only increase the lengths peaceful demonstrators will go to ... lengths of sentence and sacrifice.

As scientists our job is to research, gather data, publish results, disseminate information and write recommendations. We have the facts, the statistics and the terrifying graphs. I still believe that nonviolent civil disobedience is the duty of all those in a position to take action, which is why I will continue to take a stand against the biggest crises to ever face humanity. Not just for our species, but for all life on Earth.

In the words of Albert Einstein: 'The world will not be destroyed by those who do evil, but by those who watch them without doing anything.'

BUSES, CLIMATE DENIERS AND THE TEAM OF CLIMATE LEGENDS

Chris Jones, MA

Chris is a chemist who has taught science for nearly twenty years in ten different schools. Chris is now a climate activist, podcaster, science tutor and researcher into solutions for climate change.

You get a beautiful and terrible slice of the country when you travel around on a set of buses with some legendary activists, talking to people about the climate crisis. Picture this: a 6 foot, 2 inch-tall, mixed-race man with dreadlocks, a lab coat and no shoes shocking people with science and calming them with hope, facilitating change all over the country in the company of an amazing group of humans. Oh, and there's some nudity in there for good measure, too. Later on I'll explain how I got from being a conservatively dressed science teacher and senior leader within a set of schools, working my way to headship, to the situation above. In hearing this story you may see the value of active listening and science communication with climate deniers and the importance of hope in conversations with everyone about climate change.

***

So, what made me swap stability and comfort for more than a month touring the country on a converted bus? I spent many years aiming to be the best science teacher I could possibly be. Having taught for two years, I started talking about climate change in a serious way in 2008 in my role as gifted and talented and science enrichment coordinator. Science out of the classroom was the order of the day, and exploring big topics like climate change was deeply enshrined in what we did. Fast forward to 2019 and I had completed my master's degree in education and was training teachers. By chance, one day I bumped into some of the brightest school students that I ever taught. I had hoped that of the thousands of students that I had taught, some would aim to solve some of the world's problems, particularly this group of people. Instead, they had entered the world of finance and made themselves a pretty penny.

I reflected on this experience: when you teach about climate change for seventeen years, you are constantly reminded that the world came together to deal with ozone depletion, acid rain, lead in petrol and a whole host of other issues. But humans have not managed to solve, in more than fifty years, the problem of global warming and associated climate change. And the people in whom I had had the most faith, from some of the best schools in the country, had not tried to come up with a solution. So I decided I needed to explore other options than just teaching alone. I saw Extinction Rebellion (XR) on the news and wanted to see if activism was the way. I started to explore what this organization was doing and find out if it needed my skills in leadership, management and research. Ultimately, I decided that being a good ancestor (taking actions that have a positive effect on those that come after you) was worth more than the quick pathway to headship that I was on.

In autumn 2022, XR sent three buses on tour around the UK to spread the word about their work. When I heard about these 'rebellion buses', I felt a wave of emotions. I have British lineage and a lineage that sits outside of the UK. I love this country; I love the people in this country and I love interacting with them. I take that position even if someone's views differ from mine – we are on this little marble of a planet together and it doesn't have to heat up to the point where our societies will collapse. Therefore, I saw the potential of joining the tour; I wanted to participate, as well as review the impact of such activities. But I also feared that some XR activists might not trust data scientists.[1] Fortunately, I had already run four data projects that reviewed XR actions, and had built up enough trust that I was given the amazing opportunity to be part of the tour. I was able to travel around the country and talk to people about climate change, see what their thoughts were, what their resistances might be, and get a sense of where everyone was with regards to this existential crisis that faces us all.

***

The first bus crew I joined was a diverse collective of experienced activists led by the awesome Dolly. It was brilliant to see the faces of local XR members lighting up as we rolled up in the brightly decorated tour bus, and jumped into action signing people up, talking to deniers and creating a stir.

---

1 For example after the Facebook–Cambridge Analytica debacle, where a data company illegally used the data of thousands of people for profit.

I watched a retired doctor do some of the best outreach that I've ever seen. His manner and demeanour, knowledge of science, and position within society stopped people in their tracks. We are used to trusting personal physicians, and so it seems that we respect the opinion of a doctor who says that there's a problem. There were several teachers on that bus, too. I saw people shocked and surprised when they realized that activists come from all walks of life, and that the all-too-common narratives that climate activists are 'unemployed' or 'troubled' might be inaccurate and misleading. In watching those conversations, you could see many people experiencing cognitive dissonance, that feeling of discomfort when you're presented with stories so incompatible that one must be wrong.

People respond to cognitive dissonance in a variety of ways. Some deny the new information, while others agree with the issue but conjure a reason why it doesn't matter, saying things like, 'I think you are right about climate change but someone else will fix it.' A third group of people either simply bury their heads in the sand with comments like, 'It's too big a problem', or they change the subject at the earliest opportunity. Sound familiar?

People bring a range of emotions when faced with XR members, and quite commonly these feelings include fear and anger. In one location we were approached by several people taking the position of deniers, saying things like, 'How can areas have more snow if the world was heating up?' The words I'd learned from climate scientists at the start of my journey with XR rang in my ears and I said: 'Imagine a hot cup of coffee. Look above it and you see steam, condensation of water in cold air, right?' I extrapolated and we agreed that the sea does the same thing and if it's hotter then there is more evaporation. I continued that if that evaporation of water happens in a very cold region then the water vapour turns to ice and falls as snow, so you can have a warming planet and more snow. I could see that cognitive dissonance was falling away and we started to agree on the principles of climate change. 'Oh!' one man exclaimed. Turning a little pale, he went away and came back with a magazine promoting climate change denial. He said, 'I'm troubled that I've been lied to.' I replied, 'I'm sorry to inform you that you might have been.' He then signed up to a local XR group.

I shared these experiences with fellow activists on each of the bus tours and, unsurprisingly, similar arguments came up in various locations we visited. But now the teams were better prepared.

Overall, perhaps the most impressive outreach I saw on the entire tour was in Manchester, where the bus crew produced a combination of interesting spectacles – displays or demonstrations – to help start climate conversations, and where it was extremely busy, which meant that outreachers were being approached at least once a minute. What stood out for me was the energy and enthusiasm of the activists, and their innovative, fun conversation starters ... just like great science lessons. I observed that when music was introduced, the team's spirit was higher and more positive energy came out. Innovative approaches to outreach also led to more sign-ups, press coverage and social media engagement. For example, one experimental approach was a naked photo shoot that Dolly and several other activists did; perhaps unsurprisingly this led to an increased level of engagement and an article in the local newspaper. The outreach culminated in a people's assembly. This is a democratized process for collaborative decision-making that involves everyone having a chance to contribute and be heard in the process – generating group decisions. Following the Manchester people's assembly, I wanted to find out if some of what we had seen work well there could be replicated on another of the bus tours.

***

For the second leg of my journey, I joined the bus taking a route across the east of the country. I found a close-knit community in a crew with a more varied level of experience of outreach skills. On our journey we stayed at some pretty amazing places, from interesting individuals' homes to purpose-built communities, where we felt the warmth and sense of collective purpose. We ate, talked and partied together, trying to understand and generate ideas for our mission.

Among us was Marcus, who was arrested later that year for climbing the Queen Elizabeth II Bridge which connects to the M25 motorway around London, and unfurling a banner to raise awareness of climate change. This demonstration resulted in the temporary closure of the Dartford River Crossing, and many people view such actions as damaging to society. Marcus and I discussed the existential crisis facing us over the many evenings that we had together. We talked about crop failures, mass migration and the extreme weather that is causing flooding. We agreed that the wealthiest people believe they can simply move to wherever they please and won't be affected, so the uncomfortable truth is that climate denial and inaction serves the few, not the many.

In a little town in the north-east we encountered some of the most ingrained negative attitudes towards XR. We entered a shop and were told immediately to 'fuck off'. Following this, I hid my badges and any XR symbols, went into a different shop and proceeded to have a much more productive conversation about climate change and government inaction. That discussion led to a stranger agreeing that we found common ground. This seemed like evidence that damage done by the media, and potentially by aspects of XR's own messaging, can cause people to act fearfully or angrily upon seeing XR symbols and activists. Acknowledging this, the bus team decided to try and use hopeful messaging at the end of our outreach conversations before signing people up. We reasoned that hopeful conversations leave people with a more positive outlook and are more likely to inspire people to take action against human-caused climate change.

At this point Marcus disembarked from our crew. Little did I know that following his arrest for the protest on the Queen Elizabeth II Bridge, he would be in prison the next time many of us heard from him.

Through sharing our knowledge, reviewing what did and didn't work in each of our locations, and making use of some of the amazing techniques that I had seen in action on the first bus, our outreach skills and confidence grew. Even in the pouring rain we found ourselves able to grab people's attention and engage them. One of our best conversation starters was the question board, where people indicated how they felt about the climate crisis and how well they thought the government was responding to it. These interactive props, just like in Manchester, were helpful to start purposeful conversations about what we were doing and what actions people could take.[2] I tried various different methods used on the first bus. I had found that wearing my lab coat – a symbol many people recognize and associate with scientists – added to the success of our outreach, drawing people in and creating a more inviting spectacle. We also learned that talking to local media early in the day encouraged a broad range of people – potential activists and climate deniers alike – to come out and interact with us. We didn't shy away from conversations with our less friendly visitors, and many found they were more aligned with our vision than they had anticipated. One exclaimed, 'You aren't that bad, then!'

By the end of one of the later days on this journey, my shoes were so filled with rainwater that they refused to dry out for

2 Having a covered area where people could comfortably shelter from the rain was also a particularly helpful part of the setup!

days. So when we reached Darlington I was shoeless, dreaded and pretty exhausted, but still lab-coated and talking animatedly to the media and the wider public about our mission.

***

Science, hope, common ground, excitement, spectacle ... all help us actively and positively engage others in the climate change conversation, which is a critical precursor to getting involved in climate action. You don't have to spend a month on a bus to make a difference, but ask yourself: what *would* you consider doing to become a good ancestor?

OF SCIENCE, SYSTEMS AND SPARKLES: WHY DEMOCRACY NEEDS AN UPGRADE

Yaz Ashmawi, MPhys

Yaz holds two master's in theoretical physics from the Universities of St Andrews and Cambridge. His research was in the study of nonequilibrium systems and the physics of life, but for the last few years he has been an active campaigner. Having been involved in the strategy and action teams of Extinction Rebellion, he now coordinates Assemble, a pro-democracy movement working to update politics with a People's House in Parliament.

When I covered Sir Keir Starmer in glitter at the UK Labour Party conference in 2023, I said, 'True democracy is citizen-led.' This is because I believe the public is locked out of politics and that citizens' assemblies hand people the keys to transforming our country in the face of climate, social and ecological collapse.

Life's systems are exquisitely intricate, and what I find most wonderful about them is how improbable they are. Learning the rules and laws of physics made it seem to me like everything is predictable and that courses are set. But the deeper I looked into biology the messier it all seemed: life sings complex tunes.

I learned at university that all living things depend, moment by moment, on the reliable sum of chance interactions. Our bodies contain magnificent engines which make order out of chaos by converting noisy signals into useful work. But unlike the engines we build in everyday life, where movements are precisely calculated and identical inputs lead to identical outputs, our bodies evolved to become networks upon networks of cells and proteins built on layers upon layers of probabilistic outcomes. Whether or not a particular neuron fires or a muscle fibre contracts depends on which signalling molecules collide with which binding proteins on which cells.[1] A potentially incalculable number of factors contribute to the generation of those signals and responses. Along the way countless chemical reactions spurt and spark, cells and their components are created, damaged, supplied, destroyed, remodelled all around. Yet the body – like any ecosystem – works, accustomed to this complexity. Life finds a way.

But I also learned about how those biophysical systems can – and do – collapse. When conditions change far too quickly for life to adapt, when they become *too* unpredictable, signals get lost or misfire, resulting in failures which compound, cascade and topple into one another. Suddenly a system can dramatically

---

1 My master's work focused on incredible assemblies of proteins known as molecular motors. Surrounded by the myriad of interacting processes in their immediate environment, constantly bombarded by water molecules from all sides, these motors nevertheless channel energy forward, making the essential steps needed for muscles to function and for molecules to be transported around cells.

– and irreversibly – change its state. We see this 'tipping point' phenomenon not only in cell biology, but also in ecosystems, the climate system, and human societies and economies, too.

All of these essential systems are built on complex webs of interdependence, which gives them a certain fragility and vulnerability to tipping points. Organisms in an ecosystem both provide for and rely upon one another. For example: amphibians and algae clean up waterways; invertebrates and microbes rejuvenate soils and keep ground fertile; predators eat prey that would otherwise over-graze. It is a delicate dance and balance. In the Earth system, carbon moves between reservoirs such as the atmosphere, the oceans, living beings and geological structures, changing chemical forms as it does so. How much carbon flows through these different states, and – critically – how fast it does so, has profound impacts both within and beyond the carbon cycle.

I remember a conversation in a café with a professor at the University of Cambridge who was teaching us about the long-term history of the carbon cycle. She had explained how scientists look at important moments in geological history to try to better understand human-induced climate change today. These moments include 'The Great Dying' 250 million years back that wiped out 90 per cent of species which lived at the time, and the 'Paleocene–Eocene Thermal Maximum' 55 million years ago, which saw temperatures rise by more than 5°C for 200,000 years. Both were caused by an increase of carbon in the atmosphere – the latter came about from an injection of carbon similar to the amount humans would emit should we decide to burn our remaining fossil fuel reserves.

I asked her about something I was learning from my studies of other nonequilibrium systems: that the *rate* at which a system is pushed really matters. A very fast change can trigger 'rate-induced tipping', causing irreversible collapse much earlier than would otherwise be expected. I'd just found out that in the two mass extinction events I have mentioned, carbon was emitted into the atmosphere about *ten times slower than today.* So what did that mean for us now? Does it mean that there's no good historical analogue for the impact of today's human-caused carbon emissions? That our world is in uncharted territory, hurtling towards collapse?

I'll never forget her response. She nodded, pointedly, and said, 'Yes, it means that we seriously underestimate the sensitivity of the Earth system,' then faded into silence. I sat there waiting to be told the 'But' ... but it never came.

As the magnitude of this truth dawned on me, I felt a drop in my body – as if I was falling, as powerless to the Earth system as I would be to the force of gravity. The feeling never left me, that the people in charge, the politicians, the businesses, the leaders, *they have no idea what they are dealing with.*

If that moment never had happened, I imagine the years that followed would have found me continuing my academic research, devoted to understanding the physics of life and the secrets of probability. But instead, I found myself drawn, unexpectedly, towards very different spaces, actions and communities in attempts to contribute to revolutionary change.

For almost two years I was involved with the strategy team for Extinction Rebellion, where an important part of my role was to review the science of social change. This included the study of 'positive' tipping points which can bring about a rapid transformation in society. This was essentially the same science of systems I thought I'd left behind at Cambridge, except rather than causing collapse, these positive tipping points offer the promise of an alternative horizon. History tells us that nonviolent civil disobedience is a powerful accelerator of social change, that if enough people unite behind a shared vision, futures that may seem impossible can become inevitable.

Throughout the time I've been deepening my theoretical understanding of how societal systems change, I've been putting these ideas into action. I saw that Extinction Rebellion's mass actions on the streets of London led to Parliament declaring a climate emergency in 2019. But still there didn't seem to be enough momentum, enough political will to drive impactful climate action. So, alongside many of the other scientists in this book, I continued to participate in different forms of direct action. We replaced thousands of Tube and bus adverts to platform the science of this emergency; we occupied the Science Museum to draw attention to its endorsement of fossil-fuel companies; we pasted scientific papers to the doors of Shell's offices and covered the building in fake oil. I even walked 500 miles to Glasgow for COP26. I'm most proud of the role I played in bringing around 100,000 people together at Westminster in April 2023, united in a shared desire to protect life on Earth, united behind a shared demand that the UK government acts swiftly to end the fossil fuel era. But we still haven't seen our political leaders step up.

Democracy is a hugely powerful enabler of social and political change. But our current system of government is not fulfilling that potential. At this critical point in our history, fossil fuel lobbyists still enjoy closer relationships with political decision-makers than climate scientists do. We still see politicians prioritizing the interests of corporations and personal contacts over the needs of the people they're elected to represent. The voices represented in the corridors of power still don't reflect the diversity and experience of the billions whose lives and futures depend on the decisions made within those walls.

But we don't have to settle for a political system with these flaws and limitations. We can change the way decisions are made; we can improve and upgrade our democracy with a solution that matches the scale of the problem. A solution with much greater capability to empower the transformations that citizens – scientists and climate activists among them – have been clamouring for. We need ordinary people – across classes, ages, genders, ethnicities and abilities – to be the beating heart of government.

In the earliest known models of democracy, the people who had access to decision-making power were determined by a process of 'sortition', essentially a lottery. In the UK, we use a system like this to select juries. We trust representative groups of the public to come together, carefully consider the evidence and reach conclusions that have huge impacts on people's lives. So why not do this in our politics? Why not include people across the breadth of those affected by the choices made in Westminster, whose only objective is to take the responsibility seriously and reach solutions? When we have this option, why would we continue to settle for the structures of today's governments, where we have become so accustomed to seeing corruption, deception, prejudice and elitism influencing decisions that are essential to our well-being?

I see our current political system as the biggest barrier to the transformative action I've been pushing for ever since that moment in a small university café. Which is why I'm now focused on changing how our country is run. At the national level this means pushing for the creation of an entirely new chamber of government: a People's House – or House of Citizens[2] – consisting of an accurate snapshot of the country, everyday people paid for their time, empowered with access to experts and science, tasked to keep elected politicians accountable and deliver the policies recommended by citizens' assemblies on issues of national importance.

We know more inclusive, representative methods of political decision-making can be powerful and effective in accelerating change. The first (of many) citizens' assemblies in Ireland led

2 www.858.org.uk

to a referendum that legalized same-sex marriage, and later to abortion rights. Regional and national assemblies are becoming more widespread and influential. In Paris and Brussels, permanent assemblies have started passing laws to build public trust. In Scotland, a citizens' assembly on the future democracy of the country unearthed a deep desire for a House of Citizens. Similarly, the Climate Assembly UK in 2020[3] found that almost 90 per cent of participants agreed that we need an independent citizens' forum that monitors and ensures progress to Net Zero – including more citizens' assemblies. While this Climate Assembly had its flaws, the recommendations it created were much more ambitious and equitable than the prevailing UK policy. Nonetheless, the government was not legally obliged to follow them ... and most have been ignored. The last few years of deliberative experiments have taught us that if we want our politicians to act, we need these processes to be given real influence, that is, a central role in government and legislature.

We don't have time to wait for our political system to sort itself out. And we can't afford to. We need to use every tool we have available – from communication, to campaigning, to nonviolent protest (glitter optional!) – to press for a People's House. At the same time we need to be putting true democracy into practice within our communities. Local assemblies are already happening across the UK; people are coming together, using their collective power to decide for themselves what they want to see and building solutions to the most urgent problems of our times.

Deep down I know that no new invention or machine will provide a better solution than that most ancient of innovations: listening to and learning from one another, updating our collective understanding, collaborating and adapting. If billions of years of evolution have led to anything, surely a species which conquered

---

3 www.climateassembly.uk

the world can have the foresight to see its imminent collapse, just in time to avoid it, and pivot to chart a different course.

Life has had some close calls before. Species have survived epochs of intense volcanic activity, rising oceans, and profound changes in the composition of our atmosphere. More than two billion years ago, microscopic cyanobacteria (blue-green algae) unlocked the power of the sun by inventing photosynthesis – but this poisoned the world with the new, highly reactive molecule this process produced: oxygen.[4] The vast majority of organisms were extinguished by the waste product of what was essentially the very first energy revolution. Unlike those cyanobacteria though, we can actually see the impacts of our own fossil-fuelled energy revolution unfolding before us.

Living is miraculous and human beings are capable of incredible wisdom, compassion and love. So how can it be that we face the next mass extinction at our own hands? For me, the only explanation I can accept is that *we haven't chosen this*. But if we don't find our power to make a choice we will slide inescapably towards extinction. My hope lies in recognizing where we still have the chance to make choices. We can choose to decide together, through fair, informed, deliberative assemblies of everyday people – away from vested interests and corrupting forces – and with science at the centre. And then, at last, we will see the future transform before our eyes.

---

4 Referred to as the Great Oxygenation Event.

COMMUNITY, AGENCY AND HOPE

Shana Sullivan, MSc

Shana was born in Massachusetts and is a dual citizen of the Republic of Ireland and the United States. When she was two years old her family moved to London where she was raised. She achieved a BSc in applied physics at St. Mary's University College and an MSc in space science and engineering: space technology at University College London (UCL). She worked in the Netherlands for a year in industry before moving back to London and becoming an astronomical observatory technician and beginning a part-time PhD in astronomical instrumentation. Shana joined Extinction Rebellion in 2021, helped form the UCL Staff Climate Activist Network in 2022 and in 2023 founded the Education Climate Coalition, an organization that connects and aids collaboration between student, staff, academic and external climate action groups that operate in the educational sector.

I cannot accurately recall a time when I wasn't aware of the threat of climate change. Throughout this constant awareness, my response emotionally and materially has followed a winding path starting from acute panic, a plummet to despair and then a gentle sink into numbing passivity. It shifted suddenly, finally, with a radical energy that has transformed my life and sense of self – altering my concepts of power, agency and my sense of place and purpose in this world.

I will illustrate this journey, before activism, using a sample of three memories starting with childhood.

1

I was watching the morning news. My uniform was on and I was waiting to be walked to school. I must have been younger than eleven, still in primary school. It was the segment of the news where they looked at the newspaper headlines. Most meant next to nothing to me – except for the final one. The final newspaper they held up featured an image of an ominously backlit Earth hanging in space. The headline, something to the effect of 'Are we past the point of no return?' – a reference to the chilling fear of having reached an irreversible climate tipping point. The presenters wince, make a small noise or two of discomfort and then ... move on. Smiles return, peppy presenter voices switch back on, ushering the show briskly to the weather segment.

But as the show moves on I am left frozen. Gripped by abject terror, head spinning, breath caught: *The world is ending. It's too late – your future is on fire. You're doomed, your family is doomed. We're all going to die. We're all going to* – But wait!

Forget the end of the world for now: it's time for school. Finish your breakfast, get your shoes on. Come along.

It's little wonder I was a deeply anxious child, always thinking about the future. I wondered, why weren't the adults afraid? Why wasn't anything being done? Year after year and nothing urgent or drastic was being done. Why? Why isn't anyone doing anything? Why isn't anyone else scared? The fear drove me to obsess over the micro things in my immediate environment – recycling, turning off unused lights and worrying over the kind of car the family drove. I knew something needed doing but that I could not do it, and seemingly whoever could wasn't.

I wanted to scream but there was no place to do so – all I could do was crumple inwards.

Fear gave way to despair.

2

I'm in secondary school now, learning about population growth as part of a special 'future day' event in what would normally be a history lesson. My teacher is presenting a predicted trend chart with a line rising steeply up and insisting to me how dangerous and important this was: 'The threat of overpopulation.' I told her, bluntly, that it didn't matter. She was taken aback – probably because I was usually such a quiet and polite student – then got upset, insisting it does. But my mind was made up: 'It *doesn't* matter.' I believed deadly climate change would supersede it. Projected population growth meant nothing to me – the future felt like a lost cause anyway, a dead end. The world was ending either way. It felt like being told how many souls are on a sinking ship.

I had been despondent that entire day, worn down, anxious and depressed by the hours dedicated to talking about a future I truly believed wasn't coming.

Despite this I continued living the way I was meant to, the way I assumed I needed to, getting my GCSEs and A levels. I had dual interests – art and physics. I felt good at both but approaching university felt compelled to pick one. I chose a BSc in applied physics because I thought furthering scientific knowledge may still benefit humanity's future, whatever that would be. I distinctly felt indebted to the world. I was raised with great love and care, yet guiltily didn't seem to be capable of being 'happy' the way others could be happy. But maybe, I hoped, I could at least be *useful* in some tiny way.

I was good in academia – and found the stress of my work and relationships filled my head enough to distract me from and dull the constant ache of climate dread. I went into an MSc in space science and engineering. I liked instrumentation and space science. I took modules in Earth sciences and Earth-observing satellites for climate monitoring – still assuming that maybe my work could have some benefit – but I never dared to even think of dedicating my work to climate science.

It scared me too much.

My despair had festered into stark avoidance.

3

Now I'm just over the cusp of twenty – I'm hung-over, nursing a coffee at a greasy spoon and seated opposite a close friend. He's attempting to tell me about something he's read about recently. It's about the climate. It's not good news. I got angry, told him abruptly to stop. He tried to finish, and I utterly lost my temper and snapped, swore, silenced him, and demanded we change topic. I couldn't even talk about the things that scared me anymore. And I would rather shout down someone who did, even someone I cared about.

What had I learned, enduring nearly twenty years of climate anxiety?

Certainly not hope – that never entered the equation. What would give me hope? Hope is not blind faith – you need a reason for hope to grow, and the inaction that surrounded me provided only desolate soil.

What I cultivated instead was powerlessness. Not only the internal feeling, but also the performance of powerlessness. People who weren't me were the ones who could act – and they likely wouldn't – so that's that. The only individual survival mechanism I could then deploy at that point was burying my head in the sand: *I can't do anything about it and hearing about it stresses me out, so if I want to survive day-to-day I simply won't hear about it.*

I had to not only overcome the feeling of powerlessness, but also to unlearn that performance of powerlessness.

That began when, about five years later, I first began to see Extinction Rebellion on the streets of my home city of London. For quite a time these climate protests sat only in my life's periphery. I felt very positively towards them – but big life changes maintained that buzz in my mind that had so successfully pushed aside nagging voices. But as life calmed (and COVID restrictions lifted) that voice grew louder, insisting I was obliged to give *something, anything* – and finally pushed me into the movement in late 2021.

I showed up to the meeting points for marches, completely alone, still disempowered and thinking I would only be useful as a body for bulking out the crowd. I felt I could give so little, but I knew I still should give it. But I found roles holding banners and met other young people who shared my pain. I did police-station support alongside mothers and teachers who shared my fears. I found Scientists for XR and attended meetings, participated in actions and met individuals who shared my experiences – and with whom I wrote new ones.

Something important happened to me as I joined these communities and took action. It happened as I walked my first march, blocked my first road, attended my first meeting, painted my first banner and made my first challenges to the authorities in our society.

Where anxiety and fear once lived in my chest, another feeling began to take root. Where fear had been a constant mutter, another sound was rising in volume to challenge it. I discovered the things I had been missing and – once found – things I couldn't live without.

Community and Solidarity

The world is made up of people and systems in synergy. People create and hold up the systems, and are simultaneously constrained and controlled by them. We are inducted into systems, learn to perform within them, re-enact them, and therefore perpetuate them. The strength of systems comes from the collective power of the people performing them – often unconsciously, as I was[1] – but they are therefore weakened when we resist taking part in them. In this way, every single human being has the potential to transform the world through the transformation of themselves, and the creation of new systems and communities to resist existing ones.

In building movements we not only find people with whom we'll protest, we find individuals with whom we seek to build a better future. This work requires care, empathy and mutual aid. And therefore – if done right – results in care, empathy and mutual aid between those that undertake it. If we wish to create a fairer, gentler and compassionate world because that is where humans would flourish, then we flourish best in the here-and-now by creating that in our movements. And the creation of this

1 And still may be continuing to do so, participating in systems I have not even identified as such. For example, the performance of gender is a system that many of us subconsciously perform everyday through clothing selection and behaviour. This isn't as overt as, say, the system of capitalism – but on the flip side may be easier to resist, depending on the individual and their circumstances.

kind of community is precisely what is needed in the face of a present and future where we need (and will increasingly need) to look after one another better.

My anxiety has been replaced with an assurance that whatever the future holds I am part of a community that will respond with care and conviction. I have demonstrated to myself that I have the ability to be responsive to crises with a community I trust.

Agency

I believed, all my life, that the power to change the world was in someone else's hands. Someone more powerful than me. And therefore the only solution was to convince them to change – a task I couldn't fathom achieving.

I've come to a new belief that we should not need to appeal to 'power' for system change. Instead, a better world is made by uncovering and using the power we ourselves possess. While 'appealing to power' to make change may feel necessary in our current heavily hierarchical system (and I don't argue we shouldn't also apply pressure to existing power structures), we should not let that further lend legitimacy to those who make claim to that power and nor should we show approval of such structures. The creation of a system that *works* for the many requires power to be *held* by the many. Every act of civil disobedience and direct action is a demonstration of the power of the individual and communities to deny the status quo, hence deny it its power over us, and instead to create an opposing system. We channel energy away from the systems and hierarchies we oppose and place it into a new system. In a world made up of systems, that is of itself world-changing.

If we want a world of fairly distributed power, of people looking after each other, of action and care for the environment and justice for all, we must not merely request it but must enact it ourselves. A world of active love and care requires active humans, hence by taking action in our social movements we are manifesting this change in real time.

Agency lives within us all, but cannot be freed until we understand where our power lies. In my own circumstances[2] I could do this through taking direct action: placing myself where I once thought I could not be and doing the things I once thought I could not – and finding I can. I uncovered the extent of my own freedom through action, creating in me a person no longer paralysed by fear, despair or avoidance – but energized in all facets by love, resilience and agency.

Hope

A harsh truth of activism and social movements is that the fight never ends; there will never be a day you 'win'. It is a constant push for a better world, an ongoing struggle to a horizon you will never fully reach. But while there is no definitive win-state there is also no fail-state. We can always struggle onwards, and if our struggle has created activated, experienced individuals and resilient, caring and life-changing communities, then we have already made the world a better place along the way.

I do not fear our efforts will be for nothing if our demands are never met – I believe we have already changed the world. Because I am part of this world, and I am transformed.

So I can continue to fight to change the world, emboldened by this fact that I already have.

That fact, indisputable, gives me the hope I need.

2 As a person living in the Global North and with all the associated privileges of being white and economically secure. The agency we materially have and the ways we can express this agency differ tremendously depending on differing circumstances, and I am only speaking from my own (albeit narrow) experience.

GLOSSARY OF FREQUENTLY MENTIONED EVENTS AND PHRASES

Climate and Ecological Emergency – This term and others such as 'Climate and Nature Crisis' or 'Planetary Emergency' are used interchangeably and in various forms throughout these chapters to convey the interacting threats of human-caused global heating, destruction of ecosystems and extinction of species.

COPs (e.g. COP26) – 'Conferences of the Parties' of the United Nations Framework Convention on Climate Change (UNFCCC) take place annually. They bring together representatives of the world's governments and facilitate negotiations of international climate policy. COP21 in 2015 led to the Paris Agreement (see below). COP26 was held in Glasgow in 2021. A parallel series of COPs relating to the UN Convention on Biological Diversity also takes place roughly biennially in different parts of the world.

Global North and Global South – Rather than geographical descriptions, these are terms that attempt to group countries together according to their relative wealth, politics and power. The Global North includes the UK, the USA, countries in the European Union, Canada, Japan and South Korea, as well as Australia and New Zealand that are actually in the Southern Hemisphere. The Global South includes most of Africa and Latin America, as well as parts of the Middle East and Asia.

Greenwashing – When a government, business or other organization makes misleading or false claims that their products or activities are beneficial for the environment or have a meaningful effect on reducing environmental harms.

IPCC – The United Nations' Intergovernmental Panel on Climate Change is the body responsible for producing assessments and publishing reports on the science of climate change, its impacts, options to help adapt to these, and ways to address its underlying causes. The IPCC's 'Assessment Reports' aim to synthesize a large and diverse body of science once every few years, and are generally seen as one of the most authoritative sources on climate change.

IPCC Special Report on Global Warming of 1.5°C (2018) – This particular IPCC report warned that if global warming is not kept below 1.5°C above pre-industrial levels, then there will be significant exacerbation of floods, droughts, extreme weather events and destruction of ecosystems that will have catastrophic impacts. The report highlighted in detail how much worse the predicted impacts at 2°C are than those at 1.5°C.

Montreal Protocol – An international treaty that was agreed in 1987 by all members of the United Nations and several other states/unions to protect the Earth's ozone layer. The signatories agreed to phase out substances that deplete the ozone layer, such as chlorofluorocarbons (CFCs).

NGO – Non-governmental organization. These include not-for-profit national and international charities ranging from small groups to large international organizations.

Paris Agreement – A 2015 United Nations treaty to limit the global temperature increase to less than 2°C (1.5°C if possible) compared to pre-industrial levels, signed by 195 countries. This was one of the main outcomes of negotiations at the twenty-first Conference of the Parties (COP21), which took place in Paris.

XR – Extinction Rebellion. Founded in the UK in 2018, XR grew to become a global movement that uses nonviolent strategies to push for action to avoid catastrophic collapse of climate, ecological and societal systems. As a result of its first 'rebellion' in April 2019, by peacefully blocking roads in central London, XR propelled the climate and ecological crisis to the top of the news and the political agenda. Less than three weeks after the start of that April rebellion, MPs passed a motion making the UK Parliament the first in the world to declare an 'environment and climate emergency'.

SUGGESTED FURTHER READING

This is a list of books and other media that have been recommended by, or have influenced the thinking of, individual authors whose words are included in this book.

* Indicates XR Scientists have contributed to these books and media.

On the Climate and Ecological Emergency

Berhans, P. (2020) *The Best of Times, The Worst of Times.* Indigo Press.

Brannan, P. (2017) *The Ends of the World.* Harper Collins.

Clark, D. and Berners-Lee, M. (2013) *The Burning Question.* Profile Books.

Gergis, J. (2022) *Humanity's Moment: A Climate Scientist's Case for Hope.* Island Press.

Goulson, D. (2021) *Silent Earth: Averting the Insect Apocalypse.* Penguin.

Klein, N. (2014) *This Changes Everything: Capitalism vs the Climate.* Penguin.

Kolbert, E. (2014) *The Sixth Extinction: An Unnatural History.* Bloomsbury.

Kolbert, E. (2024) *H is for Hope: Climate Change from A to Z.* Oneworld Publications.

Lymbery, P. (2017) *Dead Zone: Where The Wild Things Were.* Bloomsbury.

Lynas, M. (2020) *Our Final Warning: Six Degrees of Climate Emergency.* Harper Collins.

Maslin, M. and Lewis, S. (2018) *The Human Planet: How we Created the Anthropocene.* Penguin.

McGuire, B. (2022) *Hothouse Earth: An Inhabitant's Guide.* Icon Books.

Oreskes, N. and Conway, E. (2018) *Merchants of Doubt: How a Handful of Scientists Obscured the Truth on Issues from Tobacco Smoke to Global Warming.* Bloomsbury.

Porritt, J. (2020) *Hope In Hell: A Decade to Confront the Climate Emergency.* Simon & Schuster UK.

Roberts, C. (2013) *The Ocean of Life: The Fate of Man and the Sea.* Penguin Random House.

* Thunberg, G. *et al.* (2022) *The Climate Book.* Penguin Random House.

Wallace-Wells, D. (2019) *The Uninhabitable Earth: A Story Of the Future.* Penguin.

On Psychology and Talking About Climate Change

Boghossian, P. and Lindsay, J. (2019) *How to Have Impossible Conversations: A Very Practical Guide.* Hachette, UK.

Gunther, G. (2024) *The Language of Climate Politics: Fossil-Fuel Propaganda and How to Fight It.* Oxford University Press USA

Hayhoe, K. (2021) *Saving Us: A Climate Scientist's Case for Hope and Healing in a Divided World.* Simon & Schuster.

Huntley, R. (2021) *How to Talk About Climate Change in a Way that*

Makes a Difference. Murdoch Books.
* Keal, L. (2021) *A Gift for Conversation: Let's Discuss Climate Change*. Available at www.agiftforconversation.com
Lewandowsky, S. and Cook, J. (2020) *The Conspiracy Theory Handbook*. Available at www.climatechangecommunication.org/conspiracy-theory-handbook/
Marshall, G. (2014) *Don't Even Think About It: Why Our Brains Are Wired to Ignore Climate Change*. Bloomsbury.
Sedgman, K. (2022) *On Being Unreasonable*. Faber.
Stoknes, P. E. (2015) *What We Think About When We Try Not to Think About Global Warming*. Chelsea Green.

On Transforming Our Organizations and Communities

Hopkins, R. (2011) *The Transition Companion: Making Your Community More Resilient in Uncertain Times*. Green Books.
Hopkins, R. (2013) *The Power of Just Doing Stuff: How Local Action Can Change the World*. Green Books.
Meadows, D. (2008) *Thinking in Systems*. Chelsea Green.
Project Drawdown. (2021) *Climate Solutions at Work: Unleashing Your Employee Power*. Available at www.drawdown.org/publications/climate-solutions-at-work
Verkade, T. and te Brömmelstroet, M. (2022) *Movement: How to Take Back Our Streets and Transform Our Lives*. Scribe Publications.

On Campaigning and Activism

Bhargava, D. and Luce, S. (2023) *Practical Radicals: Seven Strategies to Change the World*. The New Press.
Bond, B. and Exley, Z. (2016) *Rules for Revolutionaries: How Big Organizing Can Change Everything*. Chelsea Green.
Engler, M. and Engler, P. (2016) *This Is an Uprising: How Nonviolent*

Revolt Is Shaping the Twenty-First Century. Bold Type Books.
Fisher, D. R. (2024) *Saving Ourselves*. Colombia University Press.
Loach, M. (2023) *It's Not That Radical: Climate Action to Transform Our World*. DK.
Nakate, V. (2022) *A Bigger Picture: My Fight to Bring a New African Voice to the Climate Crisis*. Pan Macmillan.

On Looking After Ourselves

Grose, A. (2020) *A Guide to Eco-Anxiety: How to Protect the Planet and Your Mental Health*. Watkins.
Johnson, A. E. and Wilkinson, K. K. (2020) *All We Can Save: Truth, Courage and Solutions for the Climate Crisis*. One World.
Kennedy-Woodard, M. and Kennedy-Williams, P. (2022) *Turn the Tide on Climate Anxiety: Sustainable Action for Your Mental Health and the Planet*. Jessica Kingsley Publishers.
Loeb, P. (2004) *The Impossible Will Take a Little While: A Citizen's Guide to Hope in a Time of Fear*. Basic Books.
Macy, J. and Johnstone, C. (2012) *Active Hope: How to Face the Mess We're in Without Going Crazy*. New World Library.
Ray, S. (2020) *A Field Guide to Climate Anxiety: How to Keep Your Cool on a Warming Planet*. University of California Press.
Salamon, M. (2023) *Facing the Climate Emergency: How to Transform Yourself with Climate Truth, 2nd Edition*. New Society Publishers.
Wray, B. (2022) *Generation Dread*. Alfred A. Knopf Canada.

On Footprints and Individual Action

Berners-Lee, M. (2010) *How Bad are Bananas? The Carbon Footprint of Everything*. Profile Books.
Button, T. (2018) *A Life Less Throwaway: The Lost Art of Buying for Life*. Harper Thorsons.
Gale, J. (2020) *The Sustainable(ish) Living Guide: Everything*

You Need to Know to Make Small Changes That Make a Big Difference. Green Tree.

Grover, S. (2021) *We're All Climate Hypocrites Now: How Embracing Our Limitations Can Unlock the Power of a Movement*. New Society Publishers.

Kalmus, P. (2017) *Being the Change: How to Live Well and Spark a Climate Revolution*. New Society Publishers.

Nicholas, K. A. (2021) *Under the Sky We Make: How to Be Human in a Warming World*. G. P. Putnam's Sons, New York.

Wynes, S. (2019) *SOS: What You Can Do to Reduce Climate Change*. Ebury Press.

Articles

* Capstick, S. *et al.* (2022) 'Civil disobedience by scientists helps press for urgent climate action'. *Nature Climate Change* 12:773–774

* Finnerty, S. *et al.* (2024) 'Between two worlds: the scientist's dilemma in climate activism'. *Nature Climate Action* 3:1–11.

* Finnerty, S. *et al.* (2024) 'Scientists' identities shape engagement with environmental activism'. *Nature: Communications Earth & Environment* 5:240.

* Gardner, C., Cox, E. and Capstick, S. (2022) 'Extinction Rebellion scientists: why we glued ourselves to a government department'. *The Conversation*.

* Gardner, C. *et al.* (2021) 'From publications to public actions: the role of universities in facilitating academic advocacy and activism in the climate and ecological emergency'. *Frontiers in Sustainability* 2:42.

* Racimo, F. *et al.* (2022) 'The biospheric emergency calls for scientists to change tactics'. *eLife*: e83292.

* Thierry, A. *et al.* (2023) '"No Research on A Dead Planet":

preserving the socio-ecological conditions for academia'. *Frontiers in Education* 8:1237076.

* Wyatt, T. *et al.* (2024) 'Actions speak louder than words: the case for responsible scientific activism in an era of planetary emergency'. *Royal Society Open Science* 11:240411.
* See also research articles included under *Frontiers in Education*'s ongoing research topic 'Activating academia for an era of colliding crises', which is co-edited by Alison Green.

Other Media

* *Climate Literacy & Numeracy.* (2022) A 52-page zine by Susi Arnott. Available via www.susiarnott.co.uk/home/the-climate-emergency/climate-literacy-zine/
* *Emergency on Planet Earth.* (2021) A talk and guide by Dr Emily Grossman and XR Scientists – a synthesis of the most relevant science at the time it was written, fact-checked by a range of experts. Available via www.extinctionrebellion.uk/the-truth/the-emergency/
* *How to speak with your family and friends about environmental issues.* (2024) A simple guide to getting started written by Laura Thomas-Walters *et al.* Available via www.emieldelange.com/wp-content/uploads/2024/02/social-beings-how_to_brochure_digital.pdf
* *Plan Z: From Lab Coats to Handcuffs.* (2024) A documentary film by Louisa Jones and Vlad Morozov featuring many of the authors of this book. See www.plan-z-film.co.uk
* *Tipping Points* podcast. Hosted by the Grantham Institute at Imperial College London. Pete Knapp interviews scientists and other professionals about why they have become involved in environmental activism. Available via www.imperial.ac.uk/grantham/publications/podcasts/tipping-points/

ACKNOWLEDGEMENTS

Every time we join an action or protest, we go as individuals, and come away changed by the community we have formed. This book feels much the same. It started as many short accounts which have become woven together into a much wider collective story. We are thankful to all the authors who have shared their personal stories, for their openness and vulnerability, as well as for all they do in an effort to protect life on Earth. We hope readers feel the envelope of community that each of us gains from one another.

This work would not have been possible without the coordinating team of Drs Abi Perrin, Viola Ross-Smith and Caroline Vincent, who dedicated countless hours to reading, shaping and reviewing these chapters. In a non-hierarchical organization, every campaign must have a driving force, and this book is a credit to their passion and desire to share these stories in order to inspire action.

We have greatly appreciated the support of the team at Michael O'Mara Books, who have given this project their full support. We wish to thank in particular our editor Louise Dixon, and Lesley O'Mara for championing this work.

Our campaigns and protests owe their reach and impact to the unwavering support of numerous talented and generous individuals, all of whom we deeply appreciate. These supporters include artists, photographers, livestreamers, journalists, those who ensure our well-being, and many more who are always just a phone call away, ready to lend a hand.

For this book project, we would specifically like to thank six friends who have made invaluable contributions: Chris Packham, who graciously provided the foreword; Lucy Hogarth, whose illustrations appear throughout; Tristram Wyatt who proofread the text; Crispin Hughes whose photography features on the cover; Lynn Bjerke who is the scientist in protest in that image; and Hannah Woodhouse, whose insightful design advice elevates everything we do.

Beyond those in our immediate network we need to thank a huge – and growing – community of scientists, advocates and activists around the world: everybody who has been integral to the groundswell of scientist activism as part of Scientists for Extinction Rebellion, Scientist Rebellion and beyond; everybody who has stood up for science by taking action with Extinction Rebellion or any organization pushing for systemic change; everyone who has worked tirelessly 'behind the scenes' in campaigning, education, communication, policy, research and transforming our systems; everyone who has supported activists amidst hopeful highs and crushing lows.

We are all indebted to the activists around the world who are pushing for urgent change in significantly more challenging circumstances than our own, and especially to those who are experiencing huge personal consequences for their actions to protect life. Their courage and care inspires and motivates us.

Finally to you, the readers. Thank you for taking this journey with us. We hope that you feel reflective, curious and ready to stand up for the future of life on Earth. Every person, every species, every fraction of a degree[1] is worth fighting for.

Scientists for XR.

www.scientistsforxr.earth

1 of global heating avoided.